Neural Function

Neural Function

Michael Wang, Ph.D.

Associate Professor, Department of Physiology, Temple University School of Medicine, Philadelphia, Pennsylvania

Alan Freeman, Ph.D.

Professor and Chairman, Department of Physiology, Temple University School of Medicine, Philadelphia, Pennsylvania

Little, Brown and Company
Boston/Toronto

Copyright © 1987 by Michael Wang and Alan Freeman

First Edition

All rights reserved. No part of this book may be reproduced in any form or by any electronic or mechanical means, including information storage and retrieval systems, without permission in writing from the publisher, except by a reviewer who may quote brief passages in a review.

Library of Congress Catalog Card No. 86-81328

ISBN 0-316-92148-3

Printed in the United States of America

DON

Contents

Preface vii

I. Cellular Neurophysiology
1. Cellular Homeostasis 3
2. The Resting Cell Membrane Potential 9
3. The Action Potential 19
4. Synaptic Transmission 29
5. Excitation-Contraction Coupling 43

II. Sensory Physiology
6. Receptors 57
7. Cutaneous Sensation 65
8. Taste and Smell 79
9. Vision 87
10. Hearing 109

III. Motor Control
11. Overview of the Motor Control System 125
12. Alpha Motoneurons and Motor Units 129
13. Spinal Cord Reflexes 135
14. The Brain Stem 147
15. Basal Ganglia 157
16. The Cerebellum 163
17. The Cerebral Cortex 169

IV. Cerebral Physiology
18. Sleep 177
19. Cognitive Functions of the Brain 183

Glossary 189
Index 197

Preface

Neurosciences are integral to every medical education. Even so, the content and goals of courses devoted to the nervous system vary from program to program. This variety derives in part from a wealth of information in the field and in part from the number of scientific disciplines—hence, perspectives—involved in gathering this information. Biochemists, anatomists, cell biologists, and physiologists have all contributed to our accumulated knowledge; all participate in its teaching.

How the nervous system operates is often perceived by medical students as a particularly difficult discipline. We wrote NEURAL FUNCTION to teach the basic concepts of neurophysiology required of every medical student. Our text is divided into four sections covering cellular neurophysiology, sensory neurophysiology, motor control, and cerebral physiology and includes a glossary for quick reference. This organization promotes a functional approach to neurophysiology and will be of considerable benefit to first year medical students.

As teachers, and of course as former students, we know only too well the difficulties presented by courses in neuroscience. As authors, therefore, we have explained in great detail the difficult concepts and principles too often glossed over in other monographs. We use clinical examples and experiences familiar to the student to illustrate the workings of the nervous system. Other necessary but less difficult material is covered more briefly.

NEURAL FUNCTION is an accessible introduction to the physiology of the nervous system. Should students want to learn more about the course and location of specific pathways and structures, we trust that they will supplement their reading with an anatomy text. We hope that we have contributed to their knowledge and confidence to further explore neurophysiology.

M. W.
A. F.

I Cellular Neurophysiology

1 Cellular Homeostasis

The nervous system is composed of over one billion individual neurons that are responsible for tasks as simple as a reflex and as complex as the production of memories, insights, and thoughts required for an understanding of atomic physics. Although neurons differ in the morphology and biochemistry necessary to meet their individual functional requirements, they are all fundamentally alike. All neurons utilize the cell membrane to establish and maintain an intracellular ionic environment that is different from that of the extracellular fluid and to generate an electrical potential difference between the inside and outside of the cell.

This chapter examines those membrane properties of the neuron that enable it to create its unique intracellular environment. Subsequent chapters explain how membrane potentials are generated and used by neurons to transmit information, how neurons communicate with each other and with their effector organs, how sensory information is received and encoded, how the brain and spinal cord organize and control movement, how the brain alternates between sleep and wakefulness, and finally how the brain directs our emotional and cognitive behavior.

THE CELL MEMBRANE

Neurons, and all other cells, are enveloped by a cell membrane approximately 7.5 to 10.0 nm thick that separates the intracellular and extracellular fluids. Table 1-1 illustrates some of the important differences in the ionic concentrations of these two fluids. Notice, in particular, that the concentration of potassium and organic anions is much higher inside the cell, whereas the concentrations of sodium, calcium, and chloride are much higher outside the cell. These differences are created and maintained by the cell membrane.

Figure 1-1 is a diagrammatic view of the cell membrane, indicating those components required by neurons to carry out their individual physiological functions. The membrane is composed of a lipid bilayer into which a variety of proteins is inserted. Together these proteins provide the neuron with the capability to control the movement of material between the inside and the outside of the cell, to detect and respond to environmental stimuli, and to generate and utilize electrical potentials.

The lipid bilayer provides structural support for the cell. Important lipid-soluble metabolites, such as oxygen and carbon dioxide, are able to diffuse freely through the cell membrane and down their concentration gradients. The movement of other equally important materials, such as sodium, potassium, calcium, glucose, and amino acids that are not lipid soluble, is controlled by proteins inserted into the lipid bilayer.

As Figure 1-1 shows, some of these proteins form water-filled pores or channels through which water soluble materials can diffuse. Some of these channels are selective for particular ions, such as sodium or potassium, while others permit all substances small enough to fit into the channel to pass through it. In some cases, the channels are open all the time. In other cases, the channels are covered by gates that open in response to particular stimuli, such as a change in the cell membrane electrical potential or the binding of a chemical substance to the protein molecules forming the gated channel.

Membrane-bound proteins also act as enzymes, controlling metabolic processes that occur on the intracellular and extracellular faces of the membrane. Some of these enzymes, such as the sodium-potassium pump, participate in the translocation of material across the cell membrane. Others, such as adenylate cyclase, which is used to synthesize cyclic AMP, control the concentration of materials within the cell. Still other proteins facilitate the diffusion of substances across the cell membrane. Those materials, such as glucose, that rely on membrane protein carriers to carry them through the membrane are neither lipid soluble nor small enough to fit into the aqueous membrane channels.

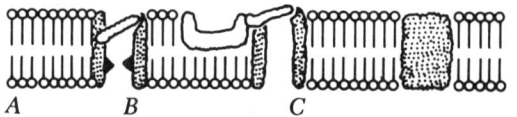

Fig. 1-1. View of a cell membrane illustrating some of the proteins that are inserted into the lipid bilayer. A represents an ion-selective channel that can be opened or closed by gate. B represents a membrane receptor that is linked to an ion channel. C represents a membrane transport protein.

TRANSLOCATION ACROSS THE CELL MEMBRANE

Most materials are translocated across the cell membrane by either diffusion or active transport.

Diffusion

Diffusion, whether simple, free diffusion, or carrier-mediated, facilitated diffusion, occurs as a result of the random movement of particles in solution. Concentration gradients cannot be established by diffusion. However, if a concentration gradient is established by an active process, the particles tend to flow from the region where they are most highly concentrated to the region where their concentration is lowest.

Fick's Law of Diffusion

Flux, or the amount of material flowing across a boundary per unit of time, is described by Fick's law of diffusion

$$\text{Flux} = \frac{D\,A}{\chi}(C_1 - C_2)$$

where D = the diffusion coefficient, A = the area through which diffusion occurs, χ = the thickness of the boundary between the regions of high and low concentrations, and C = the concentration of the material.

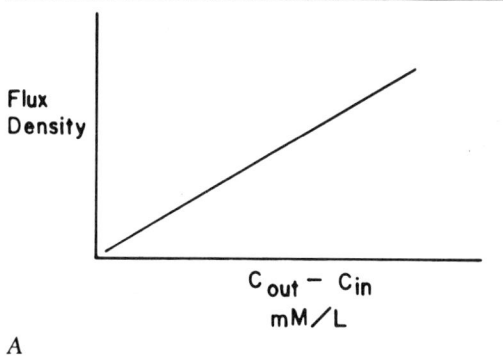

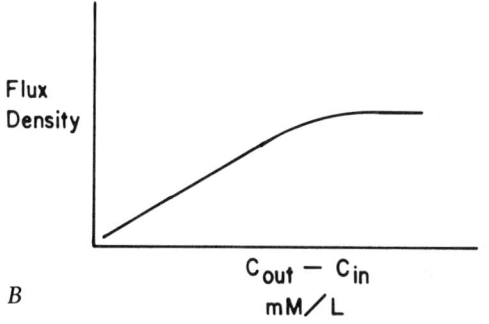

Fig. 1-2. A. Graph of the flux density crossing a membrane by diffusion as a function of the difference in concentration across the membrane. B. Graph of flux density as a function of the concentration difference when the substance is crossing the membrane by facilitated diffusion.

In the case of a cell membrane, the equation can be simplified by assuming that the thickness of all cell membranes is the same. The equation then becomes

Flux = PA $(C_1 - C_2)$

where P = the permeability coefficient.

The graph in Figure 1-2A shows the relationship between the amount of flux across a cell membrane and the concentration difference between the inside and the outside of the cell. The slope of the straight line is proportional to the permeability coefficient of the diffusing substance.

Facilitated Diffusion

Figure 1-3A shows how a membrane-bound protein participates in the translocation of glucose by facilitated diffusion. Glucose, which is more highly concentrated in the extracellular fluid, binds to the carrier protein on the outside surface of the membrane. The carrier, with its attached glucose, then diffuses to the inside surface of the membrane. Because the intracellular glucose concentration is low, glucose detaches from the carrier and enters the intracellular fluid.

The binding of glucose with the carrier obeys the law of mass action, and thus the flux across the membrane follows first-order (saturation) kinetics. This is illustrated by the graph in Figure 1-2B, depicting the relationship between flux and concentration. The flux reaches a maximum at high glucose concentrations when all of the carrier sites become filled.

Diffusion is a "downhill" process that does not require the use of free energy; that is, movement of materials by diffusion can occur only from a region of high concentration to a region of low concentration and free energy is not consumed. In contrast, energy must be used to actively transport materials from a low to a high concentration. This transport can involve the direct utilization of energy contained in ATP or can involve the energy contained in the concentration gradient of an ion. Since the concentration gradients are originally established using the energy contained in ATP, their use in translocation of material across the cell is referred to as an indirect-energy–utilizing active transport process.

Direct Active Transport

Figure 1-3B is a diagrammatic view of the *sodium-potassium pump*. This is a membrane-bound enzyme (Na-K ATPase) that is able to use the energy contained in ATP to transport sodium out of the cell while at the same time transporting potassium into the cell. The pump is responsible for establishing and maintaining the low intracellular sodium and high intracellular potassium concentrations that are shown in Table 1-1.

Although the exact mechanism by which the pump transports sodium and potassium is not known, the current view of its operation is shown in Figure 1-3B. When three sodium ions bind to the intracellular surface of the sodium-potassium pump, it is phosphorylated by an ATP molecule within the cell. The pump then undergoes a conformational change in structure that exposes the sodium ions to the extracellular fluid and weakens their binding to the pump. As a result, the sodium ions are released into the extracellular fluid. The phosphorylated enzyme then binds two potassium ions and carries them to the inside surface of the cell membrane, where the enzyme is dephosphorylated and the potassium is released into the intracellular fluid.

Each cycle of the pump thus uses the energy contained in one molecule of ATP to transport two molecules of potassium into the cell and three molecules of sodium out of the cell. Because the pump transports more positive charges out of the cell (three sodium ions) than into the cell (two potassium ions), it is electrogenic; that is, it produces an electrical potential across the cell membrane. However, the next chapter shows that the potential difference it produces is rather small (only a few millivolts at most), and thus it is not a major contributor to the cell membrane potential. The major role of the pump is to establish and maintain the sodium and potassium concentration differences across the membrane.

Indirect Active Transport

The energy contained in the sodium concentration gradient can be used to actively transport other substances against their concentration gradients. For example, amino acids are transported across the cell membrane by a sodium-dependent carrier process.

Normally, the affinity of the carrier for the amino acid is low. However, when sodium binds to the carrier, it causes an increase in the carrier's

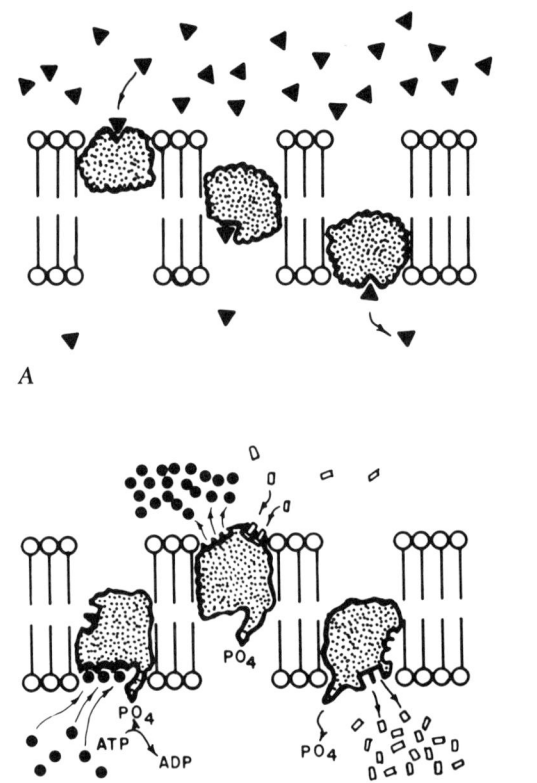

Fig. 1-3. A. Facilitated diffusion. B. The sodium-potassium active transport system. Filled circles represent sodium. Open rectangles represent potassium. A molecule of ATP provides the energy to transport the ions across the membrane against their concentration gradients.

affinity for amino acids. As a result, the amino acid can bind to the carrier even though its extracellular concentration is not particularly high. Once sodium and the amino acid are bound to the carrier, a conformational change causes both of them to be exposed to the intracellular fluid. Because the intracellular sodium concentration is low, sodium detaches from the carrier. Once sodium is removed, the carrier's affinity for the amino acid is decreased, and the amino acid enters the intracellular fluid even though it is more

Table 1-1. Differences in ionic concentrations of intracellular and extracellular fluids

Ion	Intracellular concentration (mMol/L)	Extracellular concentration (mMol/L)
Sodium (Na^+)	15	142
Potassium (K^+)	135	4
Chloride (Cl^-)	4	115
Hydrogen (H^+)	1×10^{-4}	4×10^{-5}
Calcium (Ca^{++})	1×10^{-4}	2.5
Magnesium (Mg^{++})	30	1.5
Bicarbonate (HCO_3^-)	10	24

Table 1-2. Nernst potentials of selected ions

Ion	Nernst potential (millivolts)
Sodium (Na^+)	+58
Potassium (K^+)	−92
Chloride (Cl^-)	−88
Hydrogen (H^+)	−24
Calcium (Ca^{++})	+132
Magnesium (Mg^{++})	−39
Bicarbonate (HCO_3^-)	−23

highly concentrated there than in the extracellular fluid. Similar sodium-dependent transport processes exist for a large number of substances that are important for neuronal function, such as synaptic transmitters.

Another important sodium-dependent process is involved in the control of intracellular calcium concentration. This process, called the *sodium-calcium exchange process*, uses the energy of the sodium concentration gradient to transport calcium out of the cell. Most likely, three sodium ions enter the cell for each calcium ion transported out of the cell. Like the sodium-potassium pump, the sodium-calcium exchange process is electrogenic. However, its polarity is opposite to that of the sodium-potassium pump. It makes the inside of the cell slightly positive because more positive charges (three sodium ions) enter the cell than leave it (two charges on a single calcium ion).

From the previous discussion, it is apparent that the cell membrane uses ATP to establish concentration differences between the inside and outside of the cell, and then uses the energy contained in these concentration differences to aid in the translocation of other materials necessary for normal cellular function.

Nernst Equation

Another equally important use of the energy contained in the concentration gradients that exist across the cell membrane is the establishment of the membrane potential. To explain how this is accomplished, it is necessary to convert the energy of the concentration gradient into electrical terms. This is done using an equation derived from the principles of physical chemistry, called the Nernst equation

$$E_{ion} = -\frac{RT}{zF} \ln \frac{(C)_{in}}{(C)_{out}}$$

where E = the energy of the concentration gradient expressed in millivolts, R = the natural gas constant, T = the temperature in degrees kelvin, z = the valence of the ion, F = the Faraday (a constant equal to 10^5 coulombs per equivalent that is used to convert moles to electrical charge), and $(C)_{in}$ and $(C)_{out}$ = the inside and outside concentrations of the ion involved.

If the constants are substituted into the equation it becomes

$$E_{ion} = -60 \log \frac{(C)_{in}}{(C)_{out}}$$

The electrical potential is called the Nernst or equilibrium potential and can be calculated for any ion. Table 1-2 indicates the Nernst potentials for the ions listed in Table 1-1.

The reason for referring to the Nernst potential as an equilibrium potential is made clear in Chapter 2, which deals with establishment of the resting cell membrane potential.

2 The Resting Cell Membrane Potential

THE EQUIVALENT CIRCUIT
An understanding of how neurons generate resting and action potentials can be obtained by analyzing an equivalent circuit that contains an electrical analog for each part of the neuron involved in establishing the membrane potential. Figure 2-1A is a drawing of the equivalent electrical circuit of a neuron, which consists of two batteries, two resistors (or conductors), and a capacitor. Each of these represents a constituent of the neuron that has been described in Chapter 1.

Membrane Batteries
The batteries, labeled E_{Na} and E_K, represent the energy contained in the ionic concentration gradients for sodium and potassium. As described previously, the value of the battery for each ion can be calculated using the Nernst equation.

$$E_{ion} = -60 \log \frac{(C)_{in}}{(C)_{out}}$$

Although a separate battery for each of the ions in the intracellular and extracellular fluid could be included in the equivalent circuit, only those for sodium and potassium are shown.

Membrane Conductors
The conductors, labeled G_{Na} and G_K, represent the ionic channels for sodium and potassium. The term *conductance* connotes the ease with which current flows through a conductive pathway under a given driving force. In a biological system, current is the flow of ions. Since individual ions can only move in and out of the cell through the protein-lined channels that span the membrane, membrane conductance is related to the number of open channels and the ease with which an ion can pass through each of the channels.

Membrane Channels

Although an in-depth discussion about the nature of ionic channels is beyond our scope, certain key concepts ought to be noted. The specificity of each channel is determined by a protein within the channel that is referred to as the *selectivity filter*. Another protein within the channel serves as a gate, which can be in either the open or closed position. When open, it allows ions to flow through the channel; when closed, it excludes ions from the channel.

The conductance of each channel depends on both its geometry and biophysical properties. Since each channel is only three to six angstroms wide, the single channel conductances are quite small. For example, the sodium and potassium channels have a conductance of approximately 10 picosiemens (by comparison, the element in a 100-watt light bulb is approximately one trillion times as great). By definition, a conductor having a conductance of one picosiemens (10^{-12} siemens) will allow one picoampere of current to pass through it under a driving force of one volt.

The total membrane conductance is much higher than this because there are a large number of channels inserted into the bilipid layer making up the membrane. For example, in unmyelinated axons, there are approximately 100 to 300 channels per square micron of membrane while in myelinated axons the nodes of Ranvier have a channel density of 2000 to 3000 per square micron. More important than the total number of channels in the membrane is the relative proportion of open channels selective for sodium and potassium. Thus, although the single channel conductances for sodium and potassium are approximately the same, the actual membrane conductance for these ions is different because the number of sodium and potassium channels that are open is not the same.

The gates that cover the channels determine whether the channels will be in the open or closed state. The factors (membrane potential, synaptic transmitters, or sensory stimuli) that control the opening and closing of gates vary

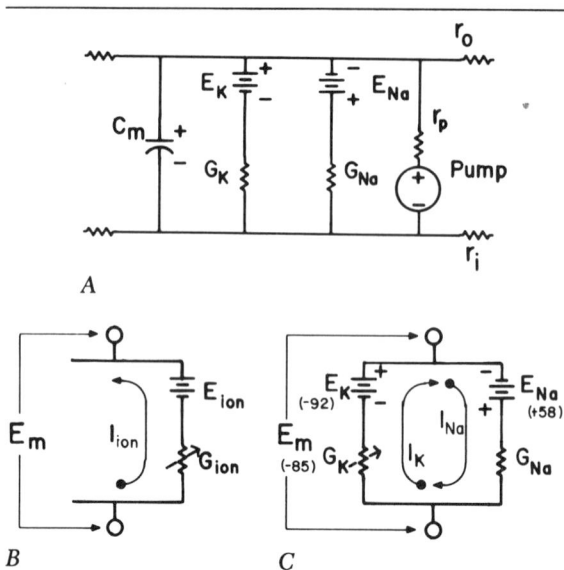

Fig. 2-1. A. An equivalent circuit representing the electrical components of the cell membrane. (The batteries, E_K and E_{Na}, represent the ionic concentration gradients; the conductances, G_K and G_{Na}, represent the ionic channels; r_o and r_i represent the extracellular and intracellular fluids; and C_m represents the membrane capacitance. The pump and the pump resistance, r_p, represent the net current driven through all of the cell membrane conductive pathways.) B. Current driven through an ionic channel by the driving force, $E_m - E_{ion}$. C. Steady-state flow of sodium and potassium current through the membrane. The values of E_m, E_K, and E_{Na} are given.

from channel to channel. This variability in membrane conductance is indicated in Figure 2-1B and C by the arrows drawn through the ionic conductors in the equivalent circuits. In the following chapters, the importance of this variation in the number of open channels is shown.

The resistors, r_o and r_i, represent the resistance of the extracellular and intracellular fluids, respectively. These resistances are not important to an understanding of how the resting membrane potential is established, but are important for describing the propagation of an action potential along the nerve membrane, which is discussed in the following chapter.

Membrane Capacitance

The capacitor, labeled C_m, represents the lipid portion of the cell membrane. Recall from the previous chapter that ions are not soluble in lipids and thus are unable to flow through the lipid portion of the membrane. Because of this, the lipid acts as a very good insulator separating two conductive media. This is, of course, the structure of a capacitor. Because the capacitor stores electrical charge, it plays a role in determining how quickly the membrane potential can be changed. The role of the membrane capacitor is discussed further in the following chapter, in considering the initiation and propagation of an action potential.

Membrane Pumps

The final component of the equivalent circuit represents the electrogenicity of the sodium-potassium pump, the sodium-calcium exchanger, and other energy-driven generators of electrical potential. Although the electrogenic pumps are included in the equivalent circuit and we will discuss how they contribute to the membrane potential, it is worth emphasizing again that the potential produced by electrogenic membrane pumps is very small and is usually ignored when considering the generation of the membrane potential.

The *membrane potential* is established by the flow of current through the membrane. When electrical charge is driven through the membrane conductors, a potential difference between the inside and outside of the cell is generated.

Ohm's Law

The relationship between the flow of current and the potential difference is given by Ohm's law

$$E = I \times R,$$
or, since $G = 1/R$,
$$I = G \times E$$

where I = the amount of current flowing in the circuit, G = the conductance, R = the resistance of the conductors in the circuit, and E = the electric potential driving current through the circuit.

By applying Ohm's law to the equivalent circuit of the neuron, we can explain how a membrane potential is established.

THE APPLICATION OF OHM'S LAW TO CELL MEMBRANES

Figure 2-1B illustrates the flow of current through a single conductive pathway of the equivalent circuit shown in Figure 2-1A. E_m represents the potential difference between the inside and outside of the cell.

The Driving Force

The electrical energy driving current through the conductor in this circuit is the difference between the membrane potential and the potential of the ionic battery. That is, driving force = $E_m - E_{ion}$.

Thus, in applying Ohm's law to biological membranes, the following equation must be used

$$I_{ion} = G_{ion} \times (E_m - E_{ion})$$

The Equilibrium State

If there is only one conductive pathway through a cell membrane (as would be the case if Figure 2-1B was the equivalent circuit for a complete cell membrane), then E_m would be equal to E_{ion}. As a result, the driving force ($E_m - E_{ion}$) would be zero and no current would flow through the membrane. This is an equilibrium state, since no free energy is required to keep the concentration gradient from changing. It is instructive to show how equilibrium is established, because this provides an intuitive understanding of how the Nernst equation is derived and also why the potential calculated with the Nernst equation is called an equilibrium potential.

Assume that a cell has the intracellular and extracellular concentrations listed in Table 1-1 for a typical cell, but that this cell is permeable only to potassium. Because potassium is more highly concentrated inside than outside the cell, it flows out of the cell through potassium-selective channels.

Electroneutrality

In order for potassium to actually enter the extracellular fluid, it must be accompanied by an anion from inside the cell or, alternatively, a cation from outside the cell must enter the intracellular fluid to replace the potassium ion flowing out. This is so because of the law of electroneutrality, which requires that a solution contain an equal number of positive and negative charges. Since there are no channels through which any other ion can flow through the membrane, potassium cannot enter the extracellular fluid.

Instead of entering the extracellular fluid, the potassium ions that manage to cross the membrane become electrostatically bound to the outside of the cell membrane. Because some potassium ions have left the intracellular fluid, there is an excess of negative charges in the intracellular fluid. In order to satisfy the requirements of electroneutrality, these ions become attached to the interior surface of the membrane. The principle of electroneutrality is not violated by the charges attached to the membrane because they are not actually in solution. However, because the charges are separated by the cell membrane, a membrane potential is established, with the inside becoming negatively charged with respect to the outside.

Establishment of Equilibrium

The negative potential inside the cell continues to build up as long as there is a net outward movement of potassium ions. However, there is a limit to the amount of potassium that can leave the inside of the cell. Recall that the force driving potassium out of the cell is the difference between the membrane potential and the potential of the potassium battery, that is, $E_m - E_K$. Initially, before any potassium has left the cell, the membrane potential is zero, so the driving force is equal to the voltage of the potassium battery. However, as potassium leaves the inside of the cell and becomes attached to the exterior surface of the cell membrane, an inside negative membrane potential develops. As a result, the driving force causing potassium to flow out of the cell becomes less. Ultimately, enough potassium becomes attached to the outside surface of the cell membrane to make the membrane potential equal to the value of the potassium battery, that is, E_m becomes equal to E_K. At this point, the driving force becomes zero and the net flow of potassium out of the cell ceases.

Derivation of Nernst Equation

The free energy contained within the concentration gradient of a substance is given by

$$\text{Energy} = RT \ln \frac{K^+_{in}}{K^+_{out}}$$

The free energy driving an ion in an electric field is given by

$$\text{Energy} = zFE$$

At equilibrium, the free energy is equal to zero, so the two terms expressed above must be equal and opposite to each other. That is,

$$zFE = -RT \ln \frac{K^+_{in}}{K^+_{out}}$$

Solving this expression for E yields the Nernst equation.

Capacitive Current

Since the ions passing through the channels become attached to the membrane in a manner analogous to charge accumulating on a capacitor, the ionic current passing through the membrane is referred to as a *capacitive current*. We will not discuss the characteristics of this current, except to point out that it represents an extremely small amount of charge. This charge is so small, in fact, that the amount of potassium leaving the cell and attaching to the outside surface of the membrane does not measurably reduce the intracellular potassium concentration.

As indicated, the situation just described represents an equilibrium condition because there is no change in ionic concentration, no net flux of ions from the intracellular to the extracellular fluid, and no utilization of free energy. Thus, the potential calculated by the Nernst equation for an ion species that is permeable to the membrane must be equal to the membrane potential if an equilibrium condition is to exist for that ion. This is why the Nernst potential is called an equilibrium potential.

The Steady State

Although an appreciation of how the equilibrium state is established is helpful in understanding how membrane potentials are established, equilibrium conditions do not actually exist in biological systems. Thus, to maintain the ionic concentrations of the intracellular fluid at a constant level, free energy must be consumed. The maintenance of a constant ionic concentration with the use of free energy is referred to as a *steady-state condition*.

Figure 2-1C depicts an equivalent circuit for a neuron in a steady-state condition. The intracellular and extracellular ionic concentrations are the same as those in the neuron described by Figure 2-1A, but in this case ionic batteries and conductive pathways are present for both sodium and potassium ions.

The values of the sodium and potassium batteries are $+58$ and -92 mV, respectively. The membrane potential is -85 mV. As we will see, this value can be predicted by applying Ohm's law to the equivalent circuit shown in Figure 2-1C.

Ionic Currents

Because the resting membrane potential is not equal to either the sodium or potassium equilibrium potential, there is a driving force causing both sodium and potassium to flow through the membrane channels that are selective for them. This driving force is also referred to as an *electrochemical gradient*, since it involves both the membrane potential and the concentration gradient.

As discussed previously, the flow of ions through membrane channels can be expressed as a current whose magnitude can be calculated by Ohm's law, $I_{Na} = G_{Na}(E_m - E_{Na})$, and $I_K = G_K(E_m - E_K)$.

Almost all of the potassium ions that flow across the membrane from inside the cell are actually able to enter the extracellular fluid. The potassium entering the extracellular fluid does not violate the principle of electroneutrality, because each positive ion leaving the intracellular fluid as a potassium ion is replaced by a positive sodium ion entering the cell from the extracellular space. In addition to the ionic flow of current through the membrane, there is a small amount of current that does not flow between the intracellular and extracellular fluids. This is the capacitive current described earlier that becomes attached to the inside and outside surfaces of the cell membrane.

Charging the Ionic Batteries

If left alone, the flow of potassium and sodium through the cell membrane (down their electrochemical gradients) causes the ionic gradients to become smaller and eventually to disappear. Ultimately, as potassium leaks out of the cell, its concentration inside the cell becomes equal to its concentration outside the cell. The leakage of sodium into the cell similarly results in its intracellular concentration becoming equal to its extracellular concentration. However, the ionic gradients are prevented from changing by the sodium-potassium pump described earlier.

The pump uses energy derived from ATP to move sodium out of and potassium into the cell. The amount of sodium and potassium moved across the cell membrane by the pump exactly balances the amount of sodium and potassium that leaks through membrane channels. In other words, the flux due to the flow of electric current through membrane channels is counterbalanced by the biochemical flux of the sodium-potassium pump. As a result, the concentration gradients do not change over time and net flux (the sum of the electrical and biochemical fluxes) is zero.

$$J_{total} = J_{ionic} + J_{biochemical} = 0$$

where J = the amount of flux.

Again, it is important to appreciate that the function of the sodium-potassium pump is to maintain the steady-state concentration gradients for sodium and potassium. Thus, the pump should be viewed primarily as a battery charger.

THE TRANSFERENCE EQUATION

Ohm's law describes the relationship between voltage and ionic fluxes through membrane channels. The biochemical fluxes ($J_{biochemical}$) produced by the pump are ignored when using Ohm's law. Thus, to derive an equation for calculating the membrane potential, only the ionic fluxes (J_{ionic}) are considered.

In the steady-state condition, three sodium ions leak into the cell for every two potassium ions that leak out of the cell. For simplicity in deriving an equation for the steady-state membrane potential, we will assume that the amounts of sodium and potassium leaking through the cell membrane are equal to each other. That is, $I_{Na} + I_K = 0$, or $I_{Na} = -I_K$.

After deriving the equation, we will show how to account for the inequality between the sodium and potassium fluxes.

It might be assumed that if the sodium and potassium currents are equal and opposite, the charges would nullify each other and that the membrane potential would be zero. This is not the case. In fact, the currents are equal because the membrane potential is not zero. Recall that the ionic current for each ion is the product of its membrane conductance and driving force. Since the conductance for potassium is much higher than it is for sodium, the two currents can be equal only if the driving force for potassium is much less than the driving force for sodium. That is, the membrane potential must be fairly close to the equilibrium potential for potassium. By applying Ohm's law to the equivalent circuit depicted in Figure 2-1C, we can derive an equation that allows us to calculate the membrane potential if the equilibrium potentials and conductances for sodium and potassium are known.

Derivation of the Transference Equation

Recall that according to Ohm's law, the sodium current is $I_{Na} = G_{Na} (E_m - E_{Na})$ and the potassium current through the membrane is $I_K = G_K (E_m - E_K)$.

Using the simplifying assumption that these currents are equal and opposite to each other, $G_{Na} (V_m - E_{Na}) = - G_K (E_m - E_k)$, and solving for the membrane potential, yields

$$V_m = \frac{G_{Na}}{G_{Na} + G_K} E_{Na} + \frac{G_K}{G_{Na} + G_K} E_K$$

Concept of Transference

The terms

$$\frac{G_{Na}}{G_{Na} + G_K} \text{ and } \frac{G_K}{G_{Na} + G_K}$$

have a special meaning. Recall that the conductance for a particular ion is the product of the single channel conductance and the total number of open channels through which that ion can pass. The total conductance (G_{total}) of the membrane is equal to the sum of all the individual ionic conductances. In this case, $G_{total} = G_{Na} + G_K$. Thus, these terms are a ratio of the individual ionic conductances (G_{Na} and G_K) to the total membrane conductance (G_{total}). This ratio is referred to as the *transference* (T_{ion}) for the ion.

The transference value for an ion expresses the fraction of open channels for a particular ion compared to the total number of open channels. For example, if the transference for sodium is given as 0.1, it means that one tenth or 10% of the open membrane channels are sodium channels. If the membrane has only sodium and potassium channels, then the remaining open channels must be specific for potassium. This means that the potassium transference is equal to 0.9 or that 90% of the open membrane channels are potassium channels. The sum of all the ionic transferences must be equal to 1.0 or 100% of all the open membrane channels.

Calculating the Membrane Potential Using the Transference Equation

If, instead of using the absolute values of the conductances in the above equation, the transferences are used, the equation becomes

$V_m = T_{Na} E_{Na} + T_K E_K$,
in which $T_{Na} = G_{Na}/G_{Total}$ and $T_K = G_K/G_{Total}$.

This equation is referred to as the *transference equation* and can be used to calculate the membrane potential during steady-state conditions. In most neurons, there are about twenty times as many open potassium channels as there are open sodium channels. Thus, the transference for sodium is

$$\frac{1}{1 + 20} = 0.05$$

The transference for potassium is

$$\frac{20}{1 + 20} = 0.95$$

Substituting these values in the transference equation yields a membrane potential of $0.95 \times (-92) + 0.05 \times (+58) = -85$ mV.

If the transferences for sodium and potassium were reversed, that is, if $T_{Na} = 0.05$ and $T_K = 0.95$, the membrane potential would be $0.05 \times (-92) + 0.95 \times (+58) = +50$ mV.

If the transferences for sodium and potassium were both equal to 0.5, then the membrane potential would be equal to a value halfway between the potassium and sodium equilibrium potentials (about -17 mV).

The equivalent circuit used to derive the transference equation contained batteries for only sodium and potassium. If the membrane contains channels for other ions, their conductances must be considered in developing the transference equation. Under these conditions, the transference equation becomes $V_m = T_{Na} E_{Na} + T_K E_K + \ldots + T_{ion} E_{ion}$. In order for an ion to make a significant contribution to the cell membrane potential, it must have a relatively high conductance. Since in most resting nerve cells sodium and potassium are the only ions to have significant membrane conductances, only these ions have been included in the transference equation.

The major assumption underlying the development of the transference equation is that the sodium and potassium currents are equal to each other. However, when these currents are measured experimentally, they are found to be unequal. Instead, there is about three times as much sodium flowing into the cell as there is potassium flowing out, as stated previously. Thus, calculation of the membrane potential using the principles of Ohm's law must take into account that the sodium current is not equal to the potassium current.

Contribution of the Sodium-Potassium Pump to the Membrane Potential

A simple expression, preserving the original transference equation, can be derived by including the contribution of the sodium-potassium pump to the membrane potential. Since the amount of sodium flowing into the cell exceeds the amount of potassium flowing out, the pump must transport more sodium than potassium. And, in fact, it has been found experimentally that the pump transports three sodium ions out of the cell for every two potassium ions it pumps in. This is just the amount it would have to pump to preserve electroneutrality and maintain the potassium and sodium batteries, given the observed ratio of ionic currents cited earlier. The excess positive charge transported by the sodium-potassium pump is referred to as the *net pump current* (I_{pump}).

The potential produced by this current can be calculated using Ohm's law.

$$E_{pump} = \frac{I_{pump}}{G_{total}}$$

The actual value of the pump potential depends on the leakage of sodium and potassium through the membrane and the magnitude of the membrane conductance. If the leakage of ions through the membrane increases, the pump must increase its pumping rate, and thus the net pump current increases. Similarly, if the membrane conductance decreases, there is an increase in the value of the pump potential produced by a given pump current. Although there are cases where the pump potential is an important component of the membrane potential, its magnitude is always a small percentage of total membrane potential.

Adding the pump potential to the transference equation yields

$$V_m = T_{Na} E_{Na} + T_K E_K + E_{pump}$$
or $V_m = E_{ionic} + E_{pump}$

where E_{ionic} = the Ohm's law or passive portion of the currents represented by the transference equation, and E_{pump} = the electrogenic contribution of the active transport system to the membrane potential.

Since E_{pump} is a rather small value, it can be ignored when calculating membrane potentials using the transference equation. Thus, using the transference equation, it is possible to calculate the magnitude of the steady-state membrane potential if the equilibrium potentials and transferences for all the ions capable of passing through membrane channels are known. The remainder of this chapter discusses how changes in the ionic gradients affect the membrane potential.

EFFECT OF IONIC GRADIENTS ON THE MEMBRANE POTENTIAL
Recording the Membrane Potential

Using finely pulled glass capillary tubes as a microelectrode, it is possible to penetrate a nerve

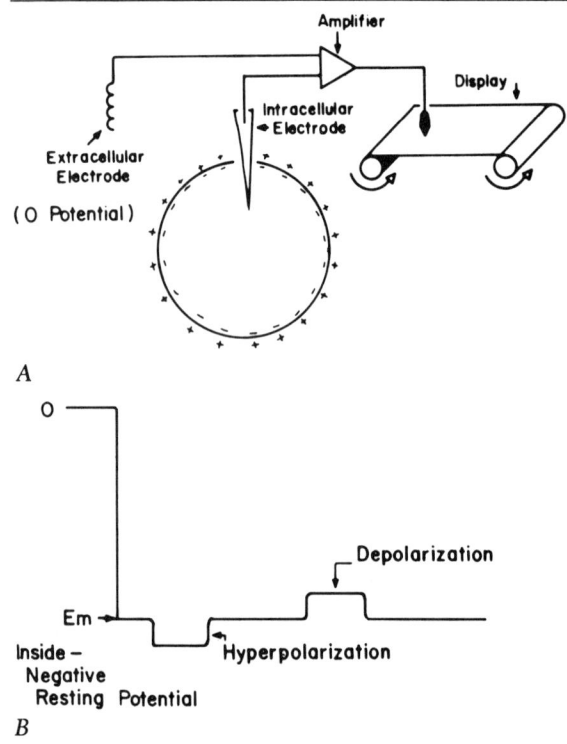

Fig. 2-2. A. Apparatus to record the intracellular membrane potential. B. The potential recorded before and after a microelectrode is inserted into the cell.

or muscle cell with minimal damage and thereby record the potential difference between the inside and outside of the cell. The electrical arrangement for this experiment is shown in Figure 2-2A.

It does not matter where the microelectrode is placed within the cell since the potential inside the cell is the same everywhere. That is, the cell interior is *isopotential*. Similarly, the potential everywhere in the extracellular fluid is the same. This potential is assumed to be zero and the potential within the cell is measured in reference to it.

Electrophysiological data are usually recorded as a function of time, using a display device such as an inkwriter or an oscilloscope. A typical resting potential as it might be recorded during an experiment is illustrated in Figure 2-2B. As long as the microelectrode is in the extracellular fluid, a potential of zero is recorded, but as soon as the microelectrode penetrates the cell membrane and enters the intracellular fluid, a negative potential is recorded. This value is the resting potential, and as long as the cell is not disturbed or injured, the value of the resting potential does not change.

The resting potential for a typical nerve cell is about -85 mV, but neuron resting potentials range from -40 to -90 mV. If the value of the resting membrane potential is made more negative, the cell is said to be *hyperpolarized*. If it is made more positive (less negative), the cell is said to be *depolarized* (Fig. 2-2B).

Under most physiological conditions, the extracellular and intracellular concentrations of sodium and potassium do not change, and therefore their equilibrium potentials do not vary. Physiological changes in the membrane potential result from changes in membrane conductance and not from changes in ionic concentrations, although in certain pathological conditions the concentration of extracellular ions may be altered.

Hyperpolarization

For example, potassium can be lost from the extracellular fluid as a side effect of a diuretic that is administered to eliminate excessive body water. Lowering extracellular potassium to 2 mMol/L causes the potassium equilibrium potential (E_K) to become -110 mV; using the transference equation, the resting membrane potential is calculated to be about -102 mV. As the graph in Figure 2-3 illustrates, the increase in membrane

Fig. 2-3. Intracellular membrane potential as a function of the extracellular potassium concentration. At lower concentrations, the membrane potential is not as high as predicted by the Nernst equation because the effects of other ions on the membrane potential become more apparent and the transference for potassium begins to decrease.

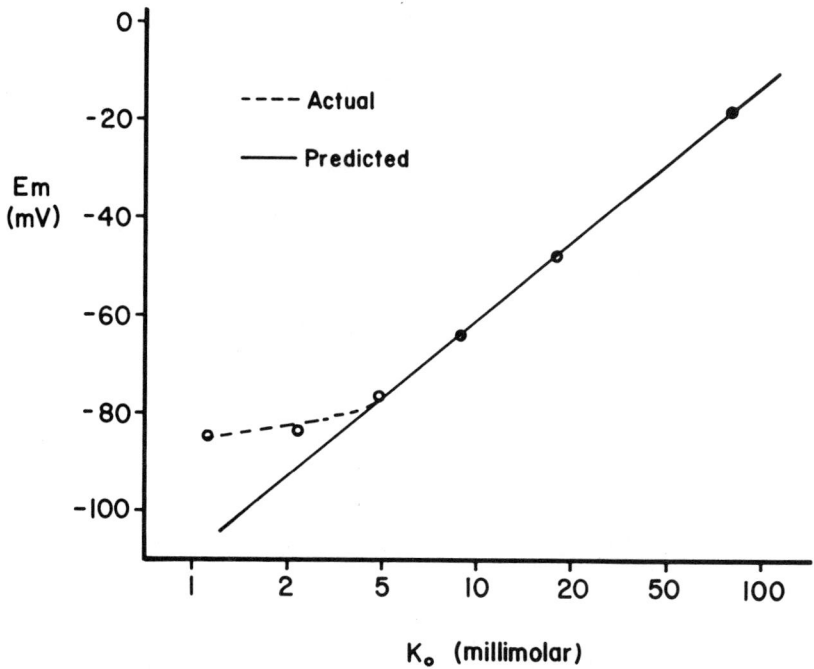

potential (membrane hyperpolarization) that results from a decrease in extracellular potassium concentration is not as great as that predicted by the transference equation because the change in potassium concentration and the consequent alteration in membrane potential cause a decrease in potassium conductance (i.e., T_K is reduced).

Depolarization

By contrast, kidney failure can cause an accumulation of extracellular potassium. If the extracellular potassium concentration increases from 4 mMol/L to 8 mMol/L, the potassium equilibrium potential becomes -75 mV. Substituting this value in the transference equation (without changing the normal values of T_{Na} and T_K) yields a membrane potential (E_m) of -68 mV. That is, increasing extracellular potassium causes the cell to depolarize.

The intracellular ionic concentrations are maintained by the sodium-potassium pump and other carrier mechanisms. Under normal physiological conditions, intracellular sodium and potassium are kept within narrow limits by the pump, but if the pump is metabolically poisoned or if adequate supplies of ATP are not available (as might occur as a result of reduced blood supply), the pump is not able to maintain internal homeostasis. Under these conditions, there is a net flux of potassium out of the cell. As a result, the potassium equilibrium potential (E_K) and consequently the membrane potential (E_m) are reduced. The cell depolarizes and normal neurological function is disturbed.

Alterations in extracellular sodium and chloride can also occur under a variety of pathological conditions, such as abnormal acid-base balance or dehydration. Because the transference of these ions is small compared to that of potassium, changes in their concentration have little effect on the resting membrane potential.

The following chapter shows how the electrophysiological behavior of nerve cells is controlled by alterations in ionic transferences.

3 The Action Potential

Figure 3-1 illustrates an oscillographic display of an action potential. As observed in Chapter 2, a microelectrode inserted into a quiescent cell records a resting potential of -85 mV. If a stimulus causes the cell to depolarize sufficiently, an action potential is generated. During the action potential, the membrane potential becomes momentarily positive and then quickly (within a few milliseconds) returns to its resting value. This chapter describes the action potential, and then discusses the mechanism by which it is generated and propagated along the axon.

CHANGES IN CONDUCTANCE RESPONSIBLE FOR THE ACTION POTENTIAL

The components of the action potential are diagrammed in Figure 3-1. The membrane is depolarized to its threshold potential by a stimulus. Once threshold is reached, the action potential begins. The first phase of the action potential is the *upstroke*. During this phase, the membrane rapidly depolarizes toward the sodium equilibrium potential. The portion of the action potential during which the membrane potential is positive is called the *overshoot*.

The next phase of the action potential is the *downstroke*. During this phase the membrane potential repolarizes toward the potassium equilibrium potential. As illustrated in Figure 3-1, the membrane potential becomes more negative than the resting potential during the downstroke. The portion of the action potential during which the membrane is hyperpolarized is called the *undershoot*.

We will explain the changes in membrane potential responsible for the action potential on two levels. First, we will use the equivalent electrical circuit presented in the previous chapter to explain how the injection of current and the subsequent changes in membrane conductance produce the action potential. Then we will describe how the ion-selective gated channels for sodium and potassium produce the changes in membrane conductance.

Passive Depolarization

Figure 3-2A is a diagram of the equivalent circuit representing the experimental arrangement used for the injection of electric current into a cell. A stimulator is connected to a microelectrode, which is inserted into the cell. When positive current is passed through the microelectrode, it causes the cell to depolarize (Fig. 3-2). Note that the membrane potential change produced by the depolarizing stimulus does not have the same time course as the stimulus. This is due to the electrical nature of the membrane which, as shown in Figure 3-2A, can be depicted as a capacitor and resistor in parallel.

The circuit in Figure 3-2A is a shortened version of the one illustrated in Figure 2-1A. It contains only those electrical components of the membrane necessary to understanding how the membrane responds passively to the injection of current. The capacitor, as usual, represents the lipid bilayer of the membrane, while the resistor represents all of the channels through which electric current can flow. When current is injected into the cell, it is divided into two paths. Some of it passes through the membrane channels (the resistor in Fig. 3-2A) and the remainder becomes attached to the membrane capacitor. Initially, most of the current goes to the capacitor, causing it to become charged. This is why the initial portion of the membrane depolarization is rapid. Then, as the membrane becomes more depolarized, more of the current passes through the resistor. This accounts for the slower rate of depolarization during the latter portion of the response to the depolarizing stimulus. Eventually, the amount of charge that the stimulus can add to the capacitor reaches a maximum and all of the current passes through the resistor. At this point, the depolarization remains at a steady value.

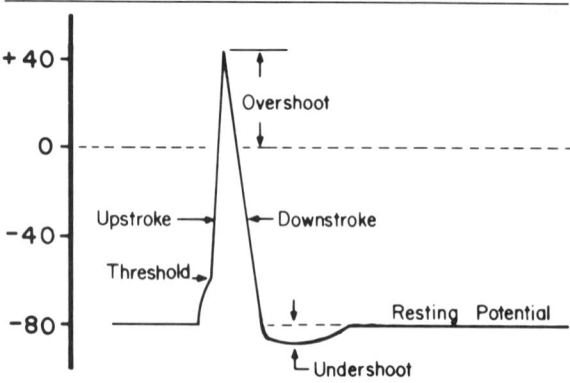

Fig. 3-1. Various components of an action potential.

The time course of the potential change caused by the application of a stimulus current can be described by the equation

$$E_t = E_{max} (1 - e^{-t/RC})$$

where E_t = the membrane potential at time, t, E_{max} = maximum voltage that the current will produce, R = the membrane resistance (the reciprocal of the conductance), and C = the membrane capacitance. The product RC is called the *time constant* of the membrane and indicates the time required for the membrane potential to reach approximately two thirds of its maximum value.

If the polarity of the injected current is reversed so that negative current is passed into the cell, the membrane becomes more negative. As illustrated in Figure 3-2B, the time course of the hyperpolarizing response is the same as that of the depolarizing response, but runs in the opposite direction. These responses are called *passive responses* because they do not cause any change in the conductance of the membrane. However, when the membrane potential is depolarized to threshold, the sodium and potassium channels actively respond to produce an action potential.

The Upstroke

The resting membrane potential of −85 mV is close to the potassium equilibrium potential be-

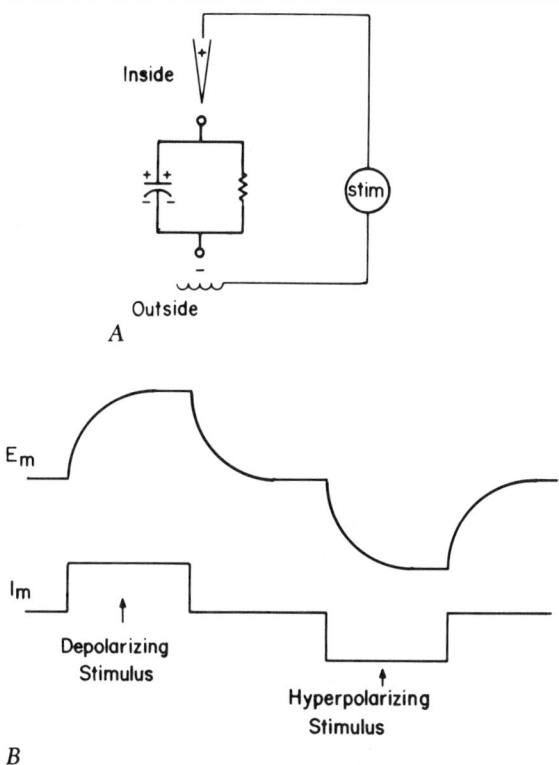

Fig. 3-2. A. Equivalent circuit of the experimental arrangement used to pass current through a cell. Some of the current entering the cell is used to charge the capacitor and the rest passes through the membrane resistor. B. Effect of the capacitive current on the time course of the change in membrane potential that occurs when depolarizing and hyperpolarizing currents are passed through the cell membrane.

cause the transference of the membrane to potassium is very high compared to its transference to sodium. This situation is shown by the circuit diagram in Figure 3-3A. The inside negativity is indicated by the negative sign on the side of the capacitor facing the inside of the membrane. The high potassium conductance is represented by the arrow pointing in the upward direction. The low sodium conductance is represented by the arrow pointing horizontally.

Recall from the previous chapter that the membrane potential can be calculated using the transference equation $E_m = E_K T_K + E_{Na} T_{Na}$ and that when the appropriate values are substituted in this equation, the resting potential is calculated to be -85 mV.

When threshold is reached, the sodium channels become activated; the gates covering the sodium channels open and the membrane conductance to sodium increases to approximately 200 times its resting value. As a result, the transference for sodium is increased from 0.05 to 0.90, while potassium transference is reduced to 0.10. This transference shift is indicated by the change in the direction of the arrows through the ionic conductances in the circuit diagram in Figure 3-3B.

The increase in sodium conductance is responsible for the upstroke of the action potential. The upstroke's magnitude can be calculated using the transference equation. Substituting the above values for sodium and potassium transferences into the transference equation, we have $E_m = (+58 \times 0.9) + (-92 \times 0.1)$ yielding a membrane potential of $E_m = +52 + -9 = +43$ mV.

During the upstroke, the membrane potential changes abruptly from its resting potential of -85 mV to an overshoot potential of $+43$ mV. The actual time course of the membrane-potential change during the upstroke depends both on how rapidly the sodium channels are activated and on the membrane time constant.

The Downstroke

As the action potential reaches the peak of its overshoot, the sodium channels are inactivated (closed), while the potassium channels are being activated (opened). These changes in ionic conductances cause the downstroke of the action potential.

The time course of the downstroke is a complex function that depends on sodium inactivation, potassium activation, and the membrane time constant. The mathematical analysis of how the membrane potential varies with time during the upstroke and downstroke is beyond the scope of this book. However, as is the case with the overshoot, the magnitude of the undershoot can be calculated easily if the sodium and potassium conductances are known. Substituting the appropriate values into the transference equation yields an undershoot potential of $E_m = +58 \times 0.02 + -92 \times 0.98 = -89$ mV.

The material presented up to this point describes the changes in ionic conductance that are responsible for the action potential. The next section explains how these changes in ionic conductance come about.

THE ROLE OF GATES IN THE PRODUCTION OF THE ACTION POTENTIAL

The sodium and potassium channels and the gates regulating the flow of ions through them are illustrated in Figure 3-4. The potassium-selective channel is the simpler of the two. It contains only one gate, called the n gate. The sodium-selective channel contains two gates, an m gate in the outer segment of the channel and an h gate in the inner portion of the channel. The m and n gates are called *activation gates* because they open during the action potential to increase the flow of ions through the sodium and potassium channels. The h gate is called the *inactivation gate* because it closes during the action potential to prevent the flow of current through the sodium channel.

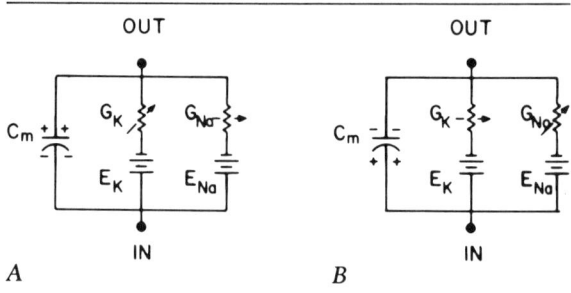

Fig. 3-3. The equivalent circuit is redrawn to indicate the change in membrane conductance that occurs during an action potential. A. The resting condition in which the potassium conductance is much higher than the sodium conductance. B. The change that occurs during the upstroke of the action potential.

The Effect of Membrane Potential on the Gates

The gates covering the sodium and potassium channels all behave in a similar fashion. They oscillate rapidly between the open and closed state in a random manner. However, because the gating molecules are charged, the probability of them being in one state or another depends on the membrane potential. When the membrane is polarized, as it is in the resting state, the gates all tend to move toward the inside of the cell. Consequently, the probability that the m and n gates are closed is high and the probability that the h

Fig. 3-4. The nature of the channels on a patch of excitable membrane. The potassium and sodium channels are shown in the resting state when the n and m gates are closed and the h gate is opened.

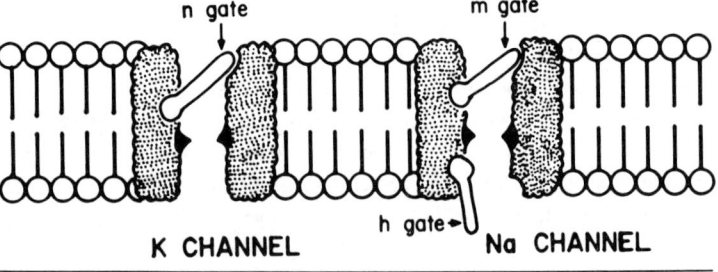

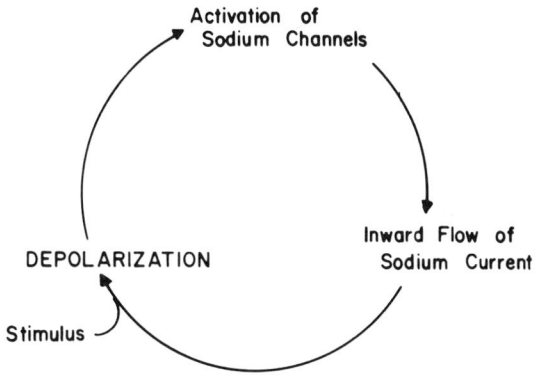

Fig. 3-5. The positive feedback cycle that occurs during the upstroke of the action potential. If the stimulus is great enough to depolarize the membrane to threshold, an action potential results.

gate is closed is low. As a result, in the resting state most of the m and n gates are closed, while most of the h gates are open.

When the membrane is depolarized, the gates move toward the outside of the cell, and thus there is an increased probability that the m and n gates will be in the open state and that the h gate will be in the closed state. Despite these similarities in the behavior of the gates in response to a change in membrane potential, there is one important difference.

The Effect of Time on the Gates

When the membrane potential changes, the gates do not all respond at the same rate. The m gates respond to a change in membrane potential much faster than the h gates, which in turn respond faster than the n gates. This means that when the membrane is depolarized, there is a sudden increase in sodium conductance due to the large number of sodium channels that are activated by the opening of m gates. This is followed by a decrease in sodium conductance as increasing numbers of h gates close to inactivate the sodium channels. Finally, the n gates begin to open the potassium channels increasing the potassium conductance. The time and voltage dependency of the gates covering the sodium and potassium channels can be used to explain the characteristics of the action potential wave form described earlier.

Activation of the Sodium Channel

In the resting state, most of the sodium and potassium channels are closed. However, there are about 20 times as many open potassium channels as there are sodium channels. For this reason, the resting membrane has a higher potassium transference and a resting potential that is near the potassium equilibrium potential. When a stimulus depolarizes the membrane to threshold, the number of m gates opening increases, and thus there is an increase in sodium conductance. This is referred to as *activation of the sodium channels*.

Regenerative (Positive Feedback) Response

The increase in sodium conductance allows sodium to enter the cell down its electrochemical gradient (or driving force) and causes the membrane to be further depolarized. This causes a further increase in the number of m gates opening and a further increase in sodium conductance. As illustrated in Figure 3-5, this cycle of sodium activation, inward flow of sodium current, and depolarization is a *regenerative* or *positive feedback response*. It accounts for the upstroke of the action potential.

Theoretically, this self-regenerative cycle could repeat itself until the membrane potential became equal to the sodium equilibrium potential. At this point, the driving force ($E_m - E_{Na}$) for sodium entry into the cell would become zero and the sodium current would cease. However, the membrane potential does not reach the sodium equilibrium potential due to the action of the h and n gates, which are also affected by the changes in membrane potential.

Inactivation of the Sodium Channel

Soon after the m gates begin to open in response to the depolarization of the membrane, the h gates begin to close, preventing the flow of sodium through the sodium channels; membrane depolarization, which opens or activates sodium channels, is thus also responsible for closing or inactivating them. The resulting decrease in sodium conductance prevents the overshoot from reaching the sodium equilibrium potential.

Activation of Potassium Channels

The n gates are the last to respond to the change in membrane potential. However, when they open to activate the potassium channels, potassium current flows out of the cell, causing the membrane to repolarize. Enough n gates are opened during the action potential to increase the potassium conductance many times its resting value. As a result, the membrane hyperpolarizes at the end of the action potential to produce the undershoot.

All-or-None Response

Once the membrane is depolarized to threshold, it undergoes a stereotypic change in membrane potential that is almost always the same. This is referred to as an *all-or-none response* and implies that all action potentials have the same waveshape regardless of the stimulus strength used to elicit them. The action potential behaves in this fashion because the gates regulating the sodium and potassium channels respond in a predictable fashion once threshold is reached.

Threshold is achieved when enough sodium channels are opened to elicit the positive feedback response. Once this occurs, the stimulus can be removed without affecting the upstroke of the action potential. For most neurons, the threshold potential is approximately -65 mV. However, this value depends on the rate of membrane depolarization.

The Local Response

Normally, when a stimulus depolarizes a membrane to threshold, the number of sodium channels activated is sufficient to initiate the regenerative or positive feedback response illustrated in Figure 3-5. However, if the membrane is depolarized too slowly (Figure 3-6A), sodium channels begin to inactivate before the regenerative

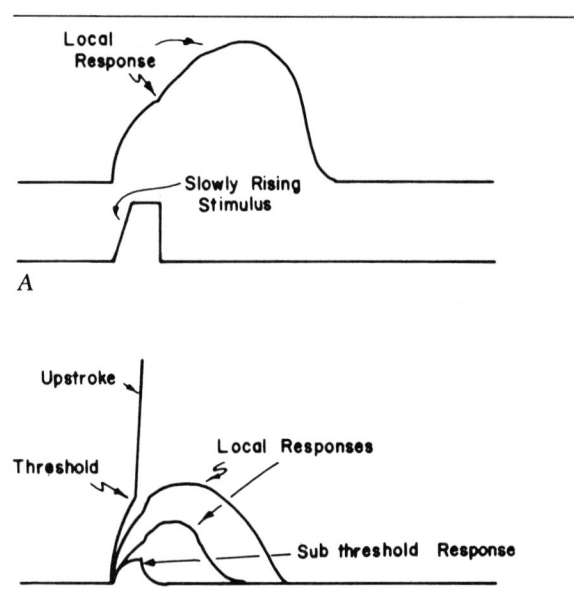

Fig. 3-6. *A. A slowly rising stimulus may not produce an action potential even though the membrane is depolarized beyond its normal threshold value. Because of the slowness of the depolarization, enough sodium channels became inactivated to prevent an action potential from developing. B. A local response can also occur when the membrane is depolarized close to threshold. In some cases, the number of activated sodium channels is insufficient to produce an all-or-none response. If the stimulus is well below threshold, no sodium channels are opened and the membrane passively depolarizes (see Fig. 3-2B).*

response occurs. As a result, the membrane returns to its resting potential without generating an action potential. The small response produced by the stimulus is called a *local response* because it is confined to the patch of membrane that was depolarized by the stimulus. This is in contrast to an action potential, which is propagated along the nerve membrane.

A local response can also occur when the stimulus applied to the membrane is just below threshold (Fig. 3-6B). Under these conditions the number of sodium channels opened is too small to initiate an action potential before sodium inactivation and potassium activation return the membrane to its resting potential. The magnitude of the local response is graded, that is, it increases in size as the strength of the stimulus increases.

The failure of a slowly applied or just subthreshold stimulus to elicit an action potential emphasizes the importance of the time- and voltage-dependent gates to the production of a normal action potential. Clearly, the sodium channels must be activated first to generate the upstroke of the action potential, but sodium inactivation and potassium activation must occur soon after to restore the membrane potential to its resting value. If these processes occur too soon, the action potential is aborted.

Refractoriness

Membrane excitability is also affected by the occurence of an action potential. If two action potentials are evoked in rapid succession, the second action potential may have an amplitude and time course that is quite different from the first (Fig. 3-7). The alterations observed in the second action potential result from a reduction in membrane excitability that always occurs during and after an action potential. The time during which there is a reduction in membrane excitability is referred to as the *refractory period*.

At the peak of the action potential, the sodium channels are inactivated by the closing of the h gates. As a result, the membrane is unable to respond to another stimulus and is completely inexcitable. This is referred to as the *absolute refractory period*.

As the membrane repolarizes, the h gates begin to open. By the time the membrane has

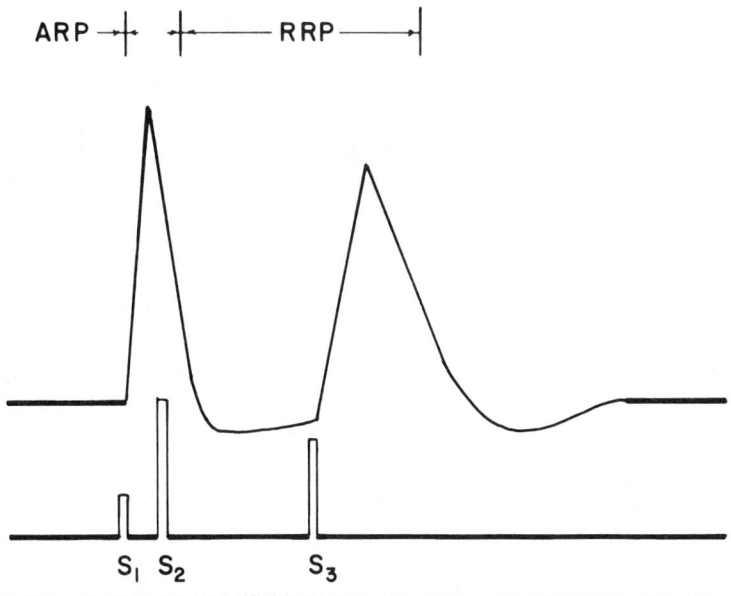

Fig. 3-7. During the absolute refractory period (ARP), an action potential cannot be elicited, no matter how great the stimulus. During the relative refractory period (RRP), an action potential can be generated by a greater than normal stimulus, but its rate of rise and magnitude is less than normal.

reached its resting potential, enough h gates will have opened to permit another action potential to be produced. When this occurs, the absolute refractory period ends, but membrane excitability does not return to normal until all the channels have returned to their resting state. This does not occur until some time after the conclusion of the absolute refractory period. The period of time during which the membrane excitability is reduced is called the *relative refractory period*. This term indicates that, although another action potential can be elicited, it is more difficult to do so. Also, the action potential produced during the relative refractory period has a lower amplitude and a slower rate of rise than a normal action potential (see Fig. 3-7).

The reduction in membrane excitability and alteration in the size and time course of the action potential during the relative refractory period are due to the condition of the channels in the period following an action potential. During this period the number of activated potassium channels is greater than normal, causing the membrane to become hyperpolarized. Since the membrane is further from threshold, a greater-than-normal stimulus must be applied to elicit an action potential. In addition, the number of inactivated sodium channels is greater than normal during this period. Thus, when an action potential is elicited, the number of sodium channels that can be activated is less than normal. This causes a reduction in the rate of rise and amplitude of the upstroke. The exact duration of the relative refractory period cannot be stated with certainty, but it roughly corresponds to the duration of the undershoot.

It is important to emphasize that the sodium-potassium pump plays no role in the generation of the action potential. All of the events described in this section, such as the threshold, the upstroke, the overshoot, the downstroke, the undershoot, and the refractory periods are due only to the properties of the voltage- and time-dependent gates.

PROPAGATION OF THE ACTION POTENTIAL

To be useful as a conveyer of information within the nervous system, the depolarization caused by an action potential must be transmitted from its point of initiation to the termination of the nerve cell. This occurs by the process of propagation or conduction of the action potential. During this process, an action potential produced by an electrical stimulus at one point on an axon causes the production of action potentials all along the axon.

Passive Spread of Current

The equivalent circuit of an axon can be used to aid in the explanation of how an action potential is propagated. Figure 3-8A is a modification of the equivalent circuit introduced in Figure 2-1A. Whereas the circuit in Figure 2-1A shows only one patch of membrane, the circuit in Figure 3-8A is extended to include several patches of membrane along the axon. Each patch is represented by a resistor (r_m) and capacitor (c_m) in parallel. The resistor represents all of the open sodium and potassium channels. Each patch of membrane is shown to be connected by a resistor representing the axoplasm, r_a, through which electric charge can flow from one excitable patch of membrane to another.

The overshoot potential of +40 mV is indicated by the positive charge on the capacitor plate facing the inside of the axon. Because the potential at this point is different from that elsewhere within the fiber, current flows between it and the rest of the axon. As current flows along the axoplasm, some of it leaves the cell through the membrane channels. The current flowing through the membrane resistors causes the cell to depolarize at that point. Because the amount of current remaining within the cell diminishes along the axon, the amount of depolarization becomes smaller as the distance between the action potential and the membrane patch becomes greater. This decrease in membrane potential with distance (illustrated by the graph in Figure

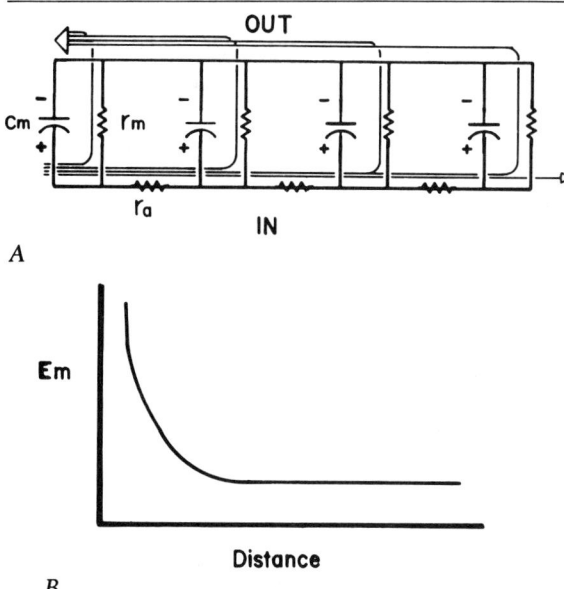

Fig. 3-8. A. *An equivalent circuit representing several contiguous patches of membrane.* B. *The decrease in membrane potential as a function of distance from the source of the current.*

3-8B) can be described by the equation $V_x = V_{max} e^{-x/\lambda}$, where V_x = the potential at distance x from the site of the action potential, V_{max} = the voltage at the site of the action potential, and λ = the space constant for the fiber. The space constant is the distance at which the voltage is reduced to approximately one third of its original value.

By analyzing the equivalent circuit, it is possible to show that the space constant for an unmyelinated axon is the square root of r_m divided by r_1. The formula for the space constant, $(r_m/r_1)^{1/2}$, indicates that the space constant increases if the membrane resistance increases. This makes intuitive sense, because an increase in membrane resistance enables less current to escape from the axon, and thus allows more current to flow to regions farther from the initial site of the action potential. Similarly, a decrease in the axoplasmic resistance, which also increases the space constant, makes it more likely for the current to flow down the axon rather than through its membrane.

Active Propagation

Each of the elements depicted in Figure 3-8A, consisting of a resistor and capacitor in parallel, represents an electrically excitable patch of membrane. That is, this patch contains the sodium and potassium channels that are required for the generation of an action potential. When the patch of membrane is depolarized to threshold by the passive spread of current into it, an action potential is initiated and the entire process begins again. Thus, the action potential generated in one patch of membrane becomes the stimulus for generating an action potential in the contiguous patch of excitable membrane. The regeneration of action potentials along the axon allows information to be rapidly spread over long distances within the nervous system. Without the regeneration of action potentials, information transfer would have to depend on the passive spread of current which, as illustrated above, diminishes rapidly along the axon.

The Effect of the Space and Time Constant on Conduction Velocity

The speed with which an action potential can be propagated depends on how rapidly action potentials can be generated at each excitable patch of membrane along the axon. One important factor influencing the conduction velocity of action potential propagation is the space constant, because it determines how much current passes through the neighboring patch of excitable membrane. Another important factor influencing the propagation velocity is the membrane capacitance, because it determines how rapidly the excitable patch of membrane is depolarized by the current passing through it. While a complete discussion of how the space constant and membrane capacitance influence the conduction velocity is beyond the scope of this book, it is important to point out that the lower the membrane capacitance and the higher the space constant, the faster the propagation velocity.

Propagation in Unmyelinated Axons

In unmyelinated axons, both the space constant and the membrane capacitance increase as fiber diameter increases. However, because the space constant increases with the square of the diameter and the capacitance is directly related to the diameter, the net effect is an increase in conduction velocity with an increase in diameter. The actual relationship is that conduction velocity in unmyelinated fibers is proportional to the square root of the diameter. Thus, the larger the axon, the more rapid the conduction velocity. However, in order to achieve reasonable conduction velocities, unmyelinated axons need to become quite large. In the squid, for example, unmyelinated fibers more than 1 mm in diameter have been found. Obviously, such an increase in fiber diameter would be impractical in human nervous systems, which contain billions of fibers.

Propagation in Myelinated Axons

In vertebrate neurons, a Schwann cell membrane covers the axon with a layer of myelin. The myelinization of nerve fibers has enabled very fast conduction velocities to be achieved without making fibers too large. For example, the largest unmyelinated fibers are about 1 micron in diameter and conduct at less than 1 m per second. By contrast, a similarly sized myelinated fiber conducts at about 6 m per second. The largest myelinated fibers in the human nervous system are approximately 20 microns in diameter and propagate at speeds of about 120 m per second.

Myelinization is able to increase propagation velocity by two mechanisms. First, it increases the membrane resistance and thus increases the space constant, and second, it decreases the membrane capacitance and thus allows the membrane to be depolarized more rapidly. Action potentials cannot be generated in membrane areas covered by myelin, they can only be initiated on the bare patches of membrane between Schwann cells, called the *nodes of Ranvier*. The nodes are about 1 to 2 mm apart. Because the action potentials appear to jump from node to node, propagation of action potentials in myelinated fibers is referred to as *saltatory conduction*.

If the myelin sheath is damaged, nerve conduction becomes impaired. In demyelinating diseases such as multiple sclerosis, conduction of nerve impulses is so impaired that normal sensory and motor tasks become impossible to perform. Less severe deficits occur following traumatic injuries to peripheral nerve fibers. Although the axons do regenerate, the myelin sheaths become thinner and the distance between nodes of Ranvier becomes less. As a result, conduction velocity is slowed.

4 Synaptic Transmission

Coordination of bodily activity by the nervous system requires nerve cells to communicate with each other and with their effector organs, the muscles and glands. This communication is accomplished by *synaptic transmission*, or the transmission of information from one cell to another across a synapse. Anatomically, the synapse consists of a presynaptic cell, usually a neuron, a synaptic space between the two cells, and a postsynaptic cell, either another neuron, a muscle, or a gland.

In a few cases within the central nervous system, synaptic transmission is accomplished by an electrical synapse in which there is a direct flow of current between the presynaptic and postsynaptic neuron. However, the most prevalent type of synaptic transmission is chemical. In chemical transmission, the presynaptic cell releases a neurotransmitter substance which, after diffusing across the synaptic space, binds to and causes a response on the postsynaptic cell. The following section briefly describes the mechanism of electrical transmission and the remainder of the chapter discusses the many varieties of chemical transmission existing within the central and peripheral nervous systems.

ELECTRICAL TRANSMISSION
Gap Junctions

Electrical transmission occurs at gap junctions formed by two cells that come in close contact with each other. The presynaptic and postsynaptic membranes are separated by a synaptic space of about 20 angstroms ($2\mu m$), which is spanned by membraneous bridges. These bridges contain a cytoplasmic channel, about 10 to 15 angstroms in diameter, that permits direct communication between the intracellular fluid of the two neurons. Thus, small particles and ions can pass directly from one cell to the other through the gap junction.

When an action potential, propagating along one cell, reaches the gap junction, current is spread passively through the cytoplasmic bridge into the other cell. If sufficient current flows to depolarize the postsynaptic cell to threshold, an action potential is generated and propagated along the postsynaptic cell. Usually, the electric current can flow in both directions through the gap junction so that a neuron can serve as both a presynaptic and postsynaptic cell.

Although gap junctions are fairly common in invertebrates and poikilothermic (cold-blooded) vertebrates, they are seldom found in the nervous systems of mammals. However, they commonly provide electrical continuity between epithelial cells and between muscle cells. This is valuable when a large group of cells needs to be rapidly and simultaneously activated. For example, in the heart and viscera where all the cells must contract together to produce efficient mechanical activity, gap junctions permit an action potential generated in one cell to rapidly activate all of the other muscle cells.

More selective and subtle communication between cells requires chemical transmission. As the following sections discuss, integration of multiple inputs, inhibition of ongoing activity, and perhaps simple forms of learning are all made possible by the sophistication of the chemical synaptic transmission process.

CHEMICAL TRANSMISSION

The simplest and best understood of all the synaptic processes is the neuromuscular synapse that occurs between an alpha motoneuron and skeletal muscle fiber. Figure 4-1 is a diagram of this synapse, showing many of its important features.

The Neuromuscular Junction

As the axon of an alpha motoneuron reaches the vicinity of a muscle, it branches many times to form numerous axon terminals. Each terminal

Fig. 4-1. The structures revealed by electronmicrographs of the skeletal muscle end plate region.

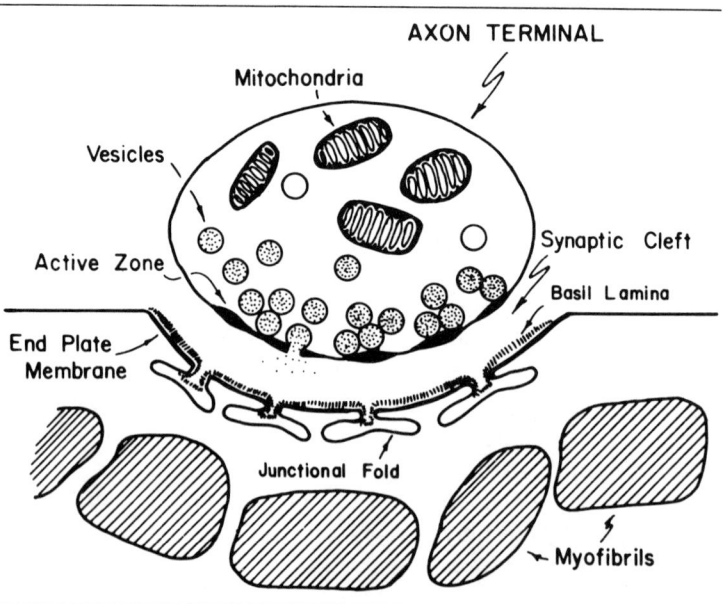

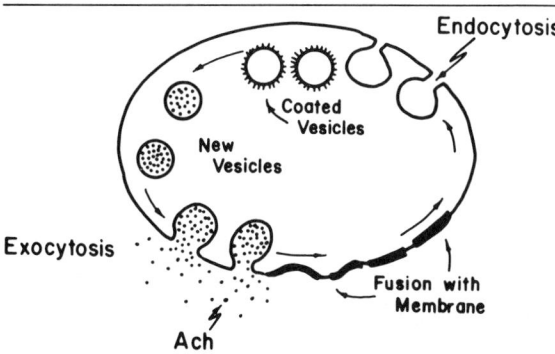

Fig. 4-2. Recycling of synaptic vesicles. After releasing their content into the extracellular space, the vesicles coalesce with the presynaptic cell membrane. Within a few minutes new vesicles are formed from invaginations of the cell membrane.

forms a synaptic junction with a single skeletal muscle fiber. A space about 500 angstroms wide separates the presynaptic (alpha motoneuron) and postsynaptic (skeletal muscle) membranes. This space is referred to as the *synaptic cleft*. The axon terminates within a small membrane invagination of the muscle fiber called the *synaptic trough* or *groove*, which is usually found in the middle of the fiber and extends for about 100 to 500 microns along the fiber. The muscle membrane forming the synaptic trough is called the *end plate*.

The Presynaptic Membrane

The presynaptic axon terminal contains thousands of small, round synaptic vesicles (see Fig. 4-1). Each is about 500 angstroms in diameter and is filled with synaptic transmitter which, in this case, is acetylcholine. The vesicles are concentrated in groups around specialized membrane structures called *dense bars*. The dense bars play an essential role in the release of synaptic transmitter from the nerve terminal. Bordering along the dense bars are rows of gated calcium channels that play an important role in the synaptic release process. The dense bars, calcium channels, and the synaptic vesicles associated with them are collectively termed the *active zone*.

The Postsynaptic Membrane

The synaptic groove is invaginated to form a series of subsynaptic or junctional folds. Embedded within the surface of these folds are acetylcholine receptor sites. These are densely packed (about 10,000 exist within 1 square micron of the postsynaptic membrane), and in addition to containing binding sites for acetylcholine, they also contain an ionic channel that is activated during the synaptic transmission process.

The synaptic cleft is filled with an amorphous structure that contains a fibrous mesh associated with the enzyme acetylcholinesterase. This enzyme breaks down acetylcholine and is necessary for terminating the postsynaptic response. The meshwork may also be involved in physically holding the synapse together.

From a physiological point of view, the synaptic transmission process may be divided into three stages: (1) release of transmitter, (2) response of the postsynaptic membrane, and (3) termination of the response. Each of these stages is explained in detail for the neuromuscular synapse, and the similarities and differences that appear in other types of synaptic transmission are indicated as they are discussed.

Excitation-Secretion Coupling

The mechanism by which synaptic transmitter is released from the presynaptic terminal is referred to as *excitation-secretion coupling*, which involves the release of synaptic transmitter from synaptic vesicles by exocytosis when an action potential is propagated into the nerve terminal. In this process, as illustrated in Figure 4-2, the membrane of the synaptic vesicle fuses with the presynaptic nerve membrane. As a result, the interior of the vesicle is momentarily exposed to the extracellular fluid, permitting the transmitter substance to diffuse out of the vesicle. Excitation-secretion coupling is initiated by the flow of calcium ions from the extracellular fluid into the nerve terminal. When the nerve terminal is depolarized by an action potential, the voltage-dependent gates that cover the calcium channels bordering the dense bars are opened, and calcium flows into the cell down its electrochemical gradient ($E_m - E_{Ca}$).

The calcium entering the presynaptic terminal during an action potential causes about 200 to 300 vesicles to release their contents into the synaptic cleft by exocytosis. Although the specific method by which calcium triggers exocytosis is not known, it is assumed to act by increasing the affinity of the vesicle for its binding site.

The amount of transmitter released by the alpha motoneuron is much greater than at other synapses, because the neuromuscular synapse is much larger than other synapses within the nervous system. As a result, each time an action potential is propagated into the nerve terminal of an alpha motoneuron, enough transmitter is released to initiate an action potential on the skeletal muscle membrane. By contrast, at other synapses within the nervous system, the neuron must fire repeatedly to cause the release of sufficient transmitter to generate an action potential on the postsynaptic membrane.

The Synaptic Vesicle Cycle

After the synaptic vesicles release their contents into the synaptic cleft, they coalesce with the presynaptic membrane (see Fig. 4-2). However, the vesicle does not remain within the neuronal membrane. Instead, the presynaptic membrane invaginates to form new vesicles. As illustrated in Figure 4-2, the invaginating membrane becomes surrounded by a clathrin coat and pinches off to form a coated vesicle.

Eventually, the clathrin coat is removed and the vesicles can then be refilled with acetylcholine and used again. Under normal physiological conditions, the recycling of vesicles is fast enough to maintain an adequate supply of vesicles within the nerve terminal. Not all of the vesicular membrane is recycled; some of it is degraded by normal cell processes and must be replaced by fresh membrane synthesized within the cell body.

Synthesis of acetylcholine takes place in the nerve terminal in close association with the vesicular membrane. The enzyme necessary for synthesis, acetylcholine transferase, is synthesized in the cell body and transported to the nerve terminal. It promotes the esterification of acytylcoenzyme A (acetyl CoA) and acetic acid to form acetylcholine. Acetyl CoA is made available as a by-product of normal cell metabolism. However, choline must be carried into the cell by a sodium-dependent transport process similar to those described in Chapter 1, Cellular Homeostasis.

Under normal physiological conditions, the quantity of vesicles can be maintained by recycling and resynthesis of acetylcholine to permit essentially continuous activity without fatigue.

The End Plate Potential

After being released from the presynaptic nerve terminal, the acetylcholine molecules diffuse across the synaptic cleft, bind to receptors on the postsynaptic membrane, and cause the postsynaptic membrane to depolarize. The depolarization produced by acetylcholine is called an *end plate potential* because it occurs on the end plate region of the muscle fiber.

The electrophysiological basis for the end plate potential is very different from that underlying an action potential. Recall that an action potential is a regenerative response produced by electrically excitable gates that goes through its entire waveform in an all-or-none manner. The end plate potential, in contrast, is produced by chemically excitable gates and does not behave in all-or-none fashion. Instead, the magnitude of the end plate potential is proportional to the amount of acetylcholine binding to the end plate receptors. The end plate potential is thus a graded response, and the number of open channels does not change with changes in membrane potential.

Another major difference between the channels responsible for the action potential and those causing the end plate potential is the ionic specificity of the channels. Two channels are required to produce the nerve action potential, a sodium channel and a potassium channel. Each of these is approximately 3 to 5 angstroms in diameter and is highly specific. In contrast, the end plate channels are about 6 angstroms in diameter and allow sodium and potassium to pass through them equally well.

The acetylcholine receptors on the end plate region of the muscle fiber are composed of two components: a binding site for acetycholine and a channel through which sodium and potassium can flow (see Fig. 1-1). When acetylcholine binds to the receptor, it causes the gate covering the ionic channel to open, allowing both sodium and potassium to flow through it. Sodium flows into the cell down its electrochemical gradient, while potassium flows out of the cell down its electrochemical gradient.

The amount of sodium and potassium flowing through the channel is given by the equations introduced in the previous chapters. Recall that $I_{Na} = G_{Na} (E_m - E_{Na})$, and that $I_K = G_K (E_m - E_K)$. These equations show that the ionic current flowing through the membrane is equal to the product of the conductance for the ion and its driving force.

The Reversal Potential

The conductance of the end plate channel opened by acetylcholine is the same for sodium and potassium ($G_{Na} = G_K$, and thus the transferences for sodium and potassium are both equal to 0.5). Since the sodium equilibrium potential ($+58$ mV) is further than the potassium equilibrium potential (-92 mV) from the membrane potential (-90 mV), more sodium current flows into the cell than potassium current flows out, and the cell depolarizes. The membrane potential resulting from the opening of the acetylcholine channel can be calculated from the transference equation developed previously

$$E_{rev} = T_{Na} E_{Na} + T_K E_K$$
$$E_{rev} = (0.5 \times -92) + (0.5 \times +58) = -17 \text{ mV}.$$

The term E_{rev} refers to the *reversal potential*. This is the potential at which the net current through the synaptic channel is zero. At a membrane potential of -17 mV, the amount of current carried into the cell by sodium ions is equal and opposite to the amount of current carried out of the cell by potassium ions, so the net current through the membrane is zero. The reversal potential is the maximum potential that can be achieved by the action of a neurotransmitter on its postsynaptic membrane.

Normally, the end plate potential never reaches the reversal potential. The actual amount of depolarization at the end plate depends on the number of acetylcholine channels open. The height of the end plate potential at a typical neuromuscular junction is only -40 mV. The membrane is not depolarized to -17 mV, because only a portion of the available acetylcholine channels are opened by the transmitter released from the presynaptic nerve terminal.

The Miniature End Plate Potential

Even in the absence of alpha motoneuron activity, there is a spontaneous discharge of acetylcholine from the nerve terminals. This results from the random attachment of vesicles to their binding sites on the nerve terminal and the consequent release of their contents by exocytosis. Because the amount of acetylcholine is approximately the same in all vesicles, the amount of transmitter released is referred to as a *quantum*. Quantal release occurs at an average rate of about 1 per second.

A single quantum contains about 10,000 acetylcholine molecules and is able to open enough postsynaptic channels to cause an end plate potential of about 1 mV. Because of their small size, the postsynaptic responses are referred to as *miniature end plate potentials* (MEPP). The physiological function, if any, of these MEPP is not known, but they are thought to play a role in maintaining the integrity of the muscle membrane because the properties of the muscle fiber are altered if its neural innervation is lost. For example, if an alpha motoneuron is damaged, all of the skeletal muscle fibers it innervates atrophy. In addition, the acetylcholine receptors that are normally concentrated on the end plate become spread out over the entire muscle membrane. Because of this latter effect of denervation, the muscle becomes much more sensitive to the action of acetylcholine. This is called *denervation supersensitivity* and is responsible for the spontaneous contractions (fibrillations) observed in muscle fibers following nerve injuries.

Inactivation of the Transmitter
The final process in synaptic transmission is the inactivation of the neurotransmitter. This step is required to prevent the transmitter from causing multiple firing of the skeletal muscle fiber. After binding to the receptor and causing the end plate channels to open, acetylcholine detaches from its binding site. If it is not rapidly removed from the vicinity of the end plate, the transmitter binds to another receptor, causing a prolonged depolarization.

Acetylcholine is inactivated by the enzyme acetylcholinesterase, which is attached to the connective tissue matrix found within the synaptic cleft. The enzyme hydrolyzes acetylcholine rapidly enough to prevent multiple firing of the skeletal muscle fibers. The choline resulting from the hydrolysis process is actively transported into the nerve terminal. It is then enzymatically recombined with acetate by choline acetyltransferase to produce a new molecule of acetylcholine.

Interestingly, the only neurotransmitter inactivated by enzymatic destruction is acetylcholine. All the other transmitters (e.g., norepinephrine, dopamine) are removed from their receptor sites by active transport into the presynaptic nerve terminal or by diffusion out of the synaptic cleft. Because these processes are slower than enzymatic destruction, the postsynaptic effects of these transmitters last longer than those of acetylcholine.

Interference with Synaptic Transmission
Synaptic transmission is a crucial process in the function of the nervous system, so it is often the target of therapeutically administered drugs and unfortunately of a number of diseases. For example, the drug hemicholinium blocks the transport of choline into the nerve terminal and thus prevents the synthesis of acetylcholine, black widow spider venom prevents the formation of new vesicles, and the botulinus toxin (from the bacterium *Clostridium botulinum*) blocks the release of acetylcholine. All of these agents ultimately result in the failure of neuromuscular transmission and thus in the paralysis of skeletal muscle.

The postsynaptic receptors can also be affected directly by a variety of drugs and disease processes. For example, the drug D-tubocurarine is capable of blocking the action of acetylcholine by binding to the acetylcholine receptor. This and similar drugs are used by anesthesiologists wishing to paralyze skeletal muscle activity during surgical procedures.

Desensitization
Another drug used to block neuromuscular transmission is succinylcholine (suxamethonium). This drug binds to the acetylcholine receptors and, like acetylcholine, causes the ionic channels to open. However, within a few seconds, the channels close and can no longer be opened by acetylcholine. This process is called *desensitization*. Although acetylcholine is also able to cause the end plate channels to desensitize, the amount of desensitization is so small that it has no physiological effect on neuromuscular transmission. The mechanism of desensitization is not well understood, but it appears to result from the inability of the gate covering the postsynaptic channel to respond to acetylcholine.

Myasthenia Gravis
Myasthenia gravis is a disease causing profound weakness of skeletal muscles. It has been discovered recently that in many cases the disease is due to an autoimmune process in which antibodies to the acetylcholine receptors are produced by the body's immunological system. The antibodies reduce the number of receptors that are able to bind to acetylcholine. As a result, the end plate

potential is reduced in size and synaptic transmission is blocked. In some cases, relief of symptoms can be achieved by thymectomy (to prevent the production of the antibodies) or by administration of immunosuppressive drugs, such as steroids.

Finally, the enzymatic destruction of acetylcholine can be affected by drugs and diseases. For example, drugs such as neostigmine, which inhibits acetylcholinesterase, can enhance the action of acetylcholine by prolonging the duration of its action. Such drugs are used to aid the recovery from the surgical use of cholinergic blocking drugs or to reduce the severity of the weakness caused by myasthenia gravis. Other types of antiacetylcholinesterases bind irreversibly to the cholinesterase molecule, permanently preventing it from hydrolyzing acetylcholine. These drugs have been used as nerve gases. Their effects result from the long duration of acetylcholine action on postsynaptic membranes of secretory glands and skeletal, cardiac, and smooth muscle.

Initiation of an Action Potential by the End Plate Potential

The ultimate purpose of synaptic transmission is to generate an action potential on the postsynaptic membrane. The end plate region of the muscle is not electrically excitable, that is, it does not have the sodium and potassium channels containing time- and voltage-dependent gates necessary to produce an action potential. However, electrically excitable channels are present on the membrane areas adjacent to the end plate. Thus, the end plate potential can act as a stimulus to depolarize the electrically excitable membrane of the muscle fiber to threshold.

Depolarization of the electrically excitable portion of the skeletal muscle membrane occurs by a process analogous to that described for the propagation of action potentials in a nerve fiber. The end plate potential creates a difference in voltage between the end plate and the region of membrane adjacent to it. Thus, current flows within the muscle fiber, depolarizing the electrically excitable portions of the membrane adjacent to the end plate. The end plate potential is normally of sufficient size to depolarize the muscle membrane to threshold and cause an action potential to be generated. In this way, whenever an action potential is propagated into the nerve terminal of an alpha motoneuron, enough acetylcholine is released to cause the generation of a skeletal muscle action potential. As the next chapter explains, the action potential causes the muscle to contract. The situation at the neuromuscular junction, in which a single postsynaptic potential is large enough to initiate an action potential, is unique within the nervous system.

At all other synapses, the postsynaptic potentials are too small to produce an action potential. As a result, the depolarization caused by many postsynaptic responses must be added together to drive the postsynaptic membrane to threshold.

CHEMICAL TRANSMISSION WITHIN THE CENTRAL NERVOUS SYSTEM

Synaptic transmission within the central nervous system is in principle no different from synaptic transmission that occurs at the neuromuscular junction. However, there are some important differences that provide the central nervous system with the ability to fine-tune the firing rate of its neuronal population. Recall that each skeletal muscle fiber receives just one synaptic ending, that the neurotransmitter is excitatory, and that each time its efferent neuron discharges, sufficient transmitter is released to cause the muscle fiber to fire. The situation is quite different in the central nervous system. These differences are illustrated by describing the synaptic processes that occur on the cell body of the alpha motoneuron.

Synaptic Transmission at the Motoneuron

The most conspicuous neuron within the spinal cord is the alpha motoneuron (Fig. 4-3). As the chapters on motor control show, the spinal cord is the final common pathway through which all components of the motor control system act to coordinate movement. There are approximately 150,000 alpha motoneurons in the human spinal cord. They vary in diameter from 25 to 100 microns and receive between 6000 and 20,000 synaptic terminals from numerous sources, including sensory axons entering the spinal cord through the dorsal roots, descending axons from the brain, and interneurons within the spinal cord. Over 80% of the soma and dendritic surface area is covered with synaptic terminals.

In contrast to the motor end plate, which can be as long as 1 mm, the synaptic terminals that form on the alpha motoneuron surface are only about 1 to 4 microns in diameter. Although the mechanism of synaptic transmitter release is the same as that governing the release of transmitter at the motor end plate, the small size of the synaptic terminal limits the number of vesicles that are able to discharge their contents into the synaptic cleft. As a result, the postsynaptic effect on the alpha motoneuron is much smaller than that observed on the end plate.

In addition, although the synaptic transmitter released by some presynaptic neurons is excitatory, transmitters released by others have an inhibitory effect on the alpha motoneuron. Thus, the firing rate of an alpha motoneuron depends on the combined effect that excitatory and inhibitory neurotransmitters have on the postsynaptic membrane. After describing the ionic mechanism underlying the excitatory and inhibitory synaptic processes, the following section explains how these two types of synapses work together to control the firing rate of the alpha motoneurons.

The Excitatory Postsynaptic Potential

The ionic mechanism underlying excitation of the alpha motoneuron is similar to that causing the end plate potential. After the neurotransmitter binds to a receptor on the postsynaptic membrane, the transmitter-receptor complex causes the opening of a membrane channel permeable to sodium and potassium. Sodium enters and potassium leaves the cell, causing the membrane to

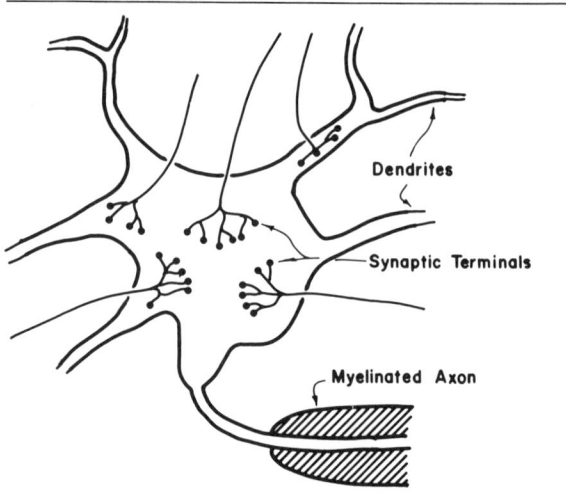

Fig. 4-3. *Alpha motoneuron. Synaptic terminals cover most of the cell body and proximal dendritic membrane.*

depolarize. The response is called an *excitatory postsynaptic potential* (EPSP).

The magnitude of the EPSP is about 5 mV (Fig. 4-4). This is much smaller than the end plate potential, because the action potential propagating along the presynaptic nerve terminal is only able to cause a few presynaptic vesicles to release their transmitters into the synaptic cleft. Thus, the number of activated excitatory channels is much less than the number of channels opened on the end plate membrane. As a result, less current flows through the alpha motoneuron membrane and the amount of depolarization is correspondingly less.

The neurotransmitters released by presynaptic nerve terminals in the spinal cord and brain have not been conclusively identified. However, a large number of substances are assumed to serve as transmitters. The most common excitatory neurotransmitters are probably the amino acids glutamate and aspartate. In addition, the eleven amino acid polypeptide, substance P, is also thought to act as an important central nervous system excitatory neurotransmitter.

As indicated previously, acetylcholine is the only transmitter inactivated by enzymatic destruction. The action of all other neurotransmitters (for example, glutamate, aspartate, substance P, and the inhibitory transmitters discussed in following sections) is terminated when the transmitter diffuses out of the synaptic cleft or is actively transported back into the presynaptic nerve terminal.

The Inhibitory Postsynaptic Potential

The type of channel activated when the inhibitory neurotransmitter binds to its receptor on the postsynaptic membrane is quite different from that activated by the excitatory transmitter. The inhibitory transmitter causes its effect on the alpha motoneuron by opening channels through which chloride can flow.

The direction of current flow depends on whether the membrane potential is positive or negative to the chloride equilibrium potential. Recall that the direction of current flow depends on the driving force ($V_m - E_{Cl}$). Thus, if the membrane potential is more positive than E_{Cl}, chloride flows down its electrochemical gradient into the cell and causes hyperpolarization. Alternatively, if the membrane potential is more negative than E_{Cl}, chloride flows out of the cell and causes the membrane to depolarize.

In most cases, the chloride equilibrium potential is more negative than the resting membrane potential, so the inhibitory neurotransmitter causes the membrane to hyperpolarize. The hyperpolarization is called the *inhibitory postsynaptic potential* (IPSP). In those cases in which the chloride equilibrium potential is equal to or more positive than the resting membrane potential, the opening of chloride channels still has an inhibitory effect on the alpha motoneuron even though the membrane does not hyperpolarize and may, in fact, depolarize. How this occurs is explained later, in considering how excitatory and inhibitory synapses combine to control the excitability of the alpha motoneuron.

The most widespread of the inhibitory neurotransmitters are the amino acids glycine and gamma-aminobutyric acid (GABA). A number of other substances, such as dopamine, norepinephrine, serotonin, and histamine, have all been strongly identified with inhibitory synaptic processes. In addition, a variety of polypeptides, particularly the morphine-like enkephalins and endorphins, are thought to be involved in inhibiting the excitability of neurons in the pain pathway of the spinal cord and brain. Other peptides, such as vasoactive intestinal polypeptide (VIP), cholecystokinin (CCK), insulin, and glucagon, that are traditionally associated with the gastrointestinal system have also been found to play a role in regulating the excitability of neurons within the central nervous system.

Initiation of an Action Potential by the EPSP

In order for the alpha motoneuron to fire an action potential, the axon hillock must be depolarized to its threshold voltage of about -65 mV. Although an action potential can be propagated over the cell body and proximal portions of the dendritic tree, the threshold for excitation is higher in these regions than at the axon hillock because of the small number of electrically excitable channels present on the cell body membrane. As indicated earlier, most of the soma and dendritic tree membrane is covered with postsynaptic chemically excitable channels.

The axon hillock is depolarized by the passive spread of current from the postsynaptic membranes of the cell body and dendrites. Recall from the discussion of the space constant in the previous chapter that the magnitude of depolarization decreases along the neuronal membrane, due to the leak of current out of the cell. Thus, the amount of depolarization appearing at the axon hillock is less than the magnitude of the EPSP generated on the cell body.

Summation

Because of the small size of the EPSP and its decrement along the cell membrane, the firing of a presynaptic neuron causes the axon hillock to depolarize by only a few millivolts. In order for threshold to be reached, the depolarization resulting from many EPSPs must be added together by the process of *summation*. Summation can occur either temporally or spatially.

Temporal summation is possible because the duration of the EPSP is longer than the refractory period of the presynaptic axon action potential. The duration of the EPSP is about 15 ms, while action potentials can be elicited in the presynaptic nerve terminal at rates of up to 1000 per second. In Figure 4-4, the size of the summated EPSP is shown as a function of the frequency of presynaptic nerve stimulation. Note that threshold is not achieved until the frequency of firing exceeds 100 per second.

Spatial summation occurs when more than one synaptic site is active at the same time. Each of the EPSPs produced on the cell membrane contribute to the depolarization of the axon hillock. Generally, threshold is achieved by a combination of temporal and spatial summation. That is, several presynaptic nerve terminals fire at high frequencies to produce enough depolarization to drive the axon hillock to its threshold of -65 mV.

Inhibition

The effects of the excitatory transmitter can be reduced by the simultaneous release of inhibitory transmitter. As illustrated in Figure 4-5, a hyperpolarizing IPSP can neutralize the depolarizing effect of an EPSP. More importantly, the presence of an IPSP can prevent the passive spread of the EPSP to the axon hillock by decreasing the space constant of the alpha motoneuron membrane.

Recall that the space constant is the distance at which the size of the EPSP falls to one third of its original value. The exact value of the space constant cannot be calculated as easily for a cell body as it can be for the nerve axon discussed in the previous chapter. However, like the axon, the magnitude of the space constant decreases as the membrane resistance decreases. When an IPSP causes

---

Fig. 4-4. *In order for an action potential to be generated on an alpha motoneuron, summation is required.*

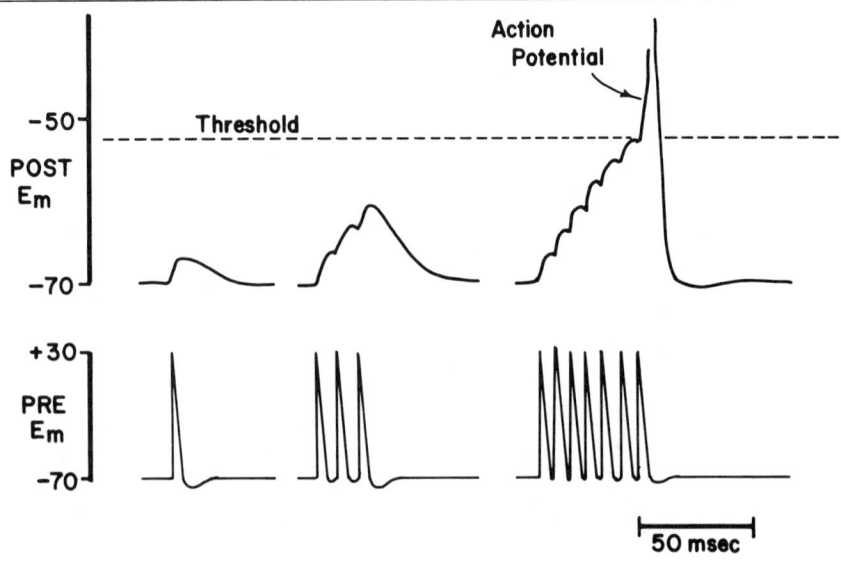

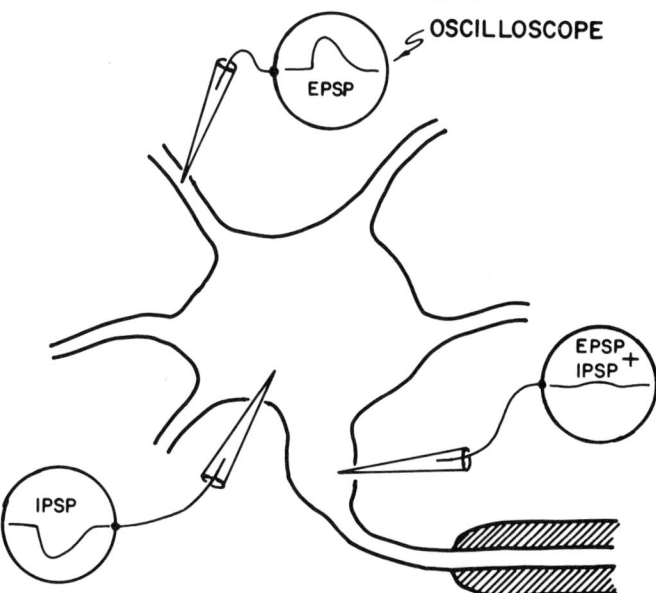

Fig. 4-5. Excitatory and inhibitory postsynaptic potentials summate at the axon hillock. If the IPSP is great enough, the alpha motoneuron can be prevented from reaching threshold.

chloride channels to open, it also causes the total membrane resistance to decrease and thus causes a decrease in the space constant. This decrease in the space constant causes a decrease in the magnitude of the EPSP reaching the axon hillock, even if the inhibitory synapse does not cause the alpha motoneuron to hyperpolarize. In general, an IPSP reduces the size of an EPSP by about 50 to 60%, regardless of its direction (hyperpolarizing or depolarizing). Obviously, for the IPSP to be most effective in reducing the size of the EPSP, it must be generated closer to the axon hillock than the EPSP. Fortunately, this is the case. Most of the excitatory synapses terminate near or on the dendrites, while the inhibitory synapses end closer to the axon hillock.

Presynaptic Inhibition

Another important mechanism used to control the firing frequency of an alpha motoneuron is called *presynaptic inhibition*. In this process, the excitatory nerve terminal is itself inhibited, so that it releases a smaller quantity of neurotransmitter. The neuronal arrangement required for presynaptic transmission to occur is illustrated in Figure 4-6. The excitatory nerve terminal is shown to receive a synaptic input. Such a synapse is called an *axoaxonic synapse* because both the presynaptic and postsynaptic membranes are axons.

The amount of transmitter released by a neuron depends on the number of synaptic vesicles that bind to their attachment sites. Recall that this depends on the amount of calcium entering the nerve terminal when it is depolarized by an action potential. If the height of the action potential is reduced, the amount of calcium entering the nerve terminal is also reduced. As a result, the amount of transmitter released from the terminal and consequently the size of the EPSP is reduced.

The transmitter (most likely GABA) released by the presynaptic neuron of the axoaxonic synapse has an inhibitory effect on the axon that normally causes an EPSP to be produced on the alpha motoneuron. As a result of this inhibition, the magnitude of the action potential invading the nerve terminal is reduced. Because the action potential is

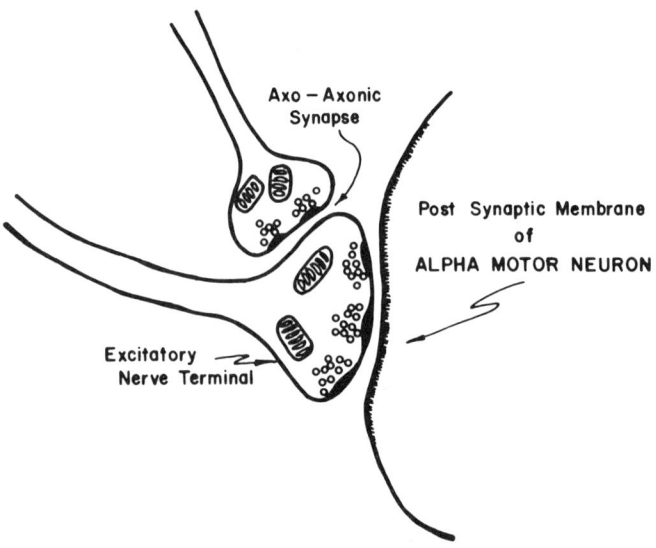

Fig. 4-6. Presynaptic inhibition is produced by an axo-axonic synapse. The transmitter released at the axo-axonic synapse reduces the magnitude of the action potential that invades the excitatory nerve terminal. As a result the size of the EPSP on the alpha motoneuron is reduced.

reduced in size, less calcium enters the nerve terminal, less neurotransmitter is released, and the EPSP is reduced in size.

Presynaptic inhibition is an important mechanism of inhibition because it permits the selective inhibition of particular excitatory inputs to the alpha motoneuron. Postsynaptic inhibition, on the other hand, acting through IPSPs, depresses the excitability of the alpha motoneuron and thus reduces the influence of all excitatory inputs to it.

CHEMICAL TRANSMISSION IN THE AUTONOMIC NERVOUS SYSTEM

The autonomic nervous system is usually subdivided into sympathetic and parasympathetic divisions. Each division is thought of as being part of a two-neuron chain, a preganglionic neuron with its cell body in the central nervous system and a postganglionic neuron with its cell body in a peripheral ganglion. The sympathetic nervous system has its preganglionic neurons in either the thoracic or lumbar spinal cord, while the parasympathetic nervous system has its preganglionic neurons in the sacral spinal cord or brain stem.

This simple description is no longer adequate for understanding the complexity of the role played by the autonomic nervous system in controlling homeostasis and behavior. For example, the enteric nervous system found within the gastrointestinal system has a richness of neuronal circuitry and a variety of neurotransmitters that rival those found within the central nervous system. However, the classical view of the autonomic nervous system defines the major sites of parasympathetic and sympathetic synaptic transmission considered in this chapter.

Autonomic Ganglia

The autonomic ganglia have been considered as simple relay stations for the transmission of information from the central nervous system to the viscera. But like most generalizations about the autonomic nervous system, this is an oversimplified view. Nonetheless, it is clear that the transmitter released by the preganglionic cells is acetylcholine, and that its effect on the postsynaptic cell membrane is very similar to the effect described for acetylcholine's action on skeletal muscle.

However, unlike the situation at the end plate of skeletal muscle, the number of quanta released at a synapse within an autonomic ganglion is not sufficient to bring the postsynaptic cell to threshold. Thus, extensive temporal and spatial summations are required to initiate an action potential in the postsynaptic neuron. As each postganglionic cell receives synaptic input from many preganglionic neurons, neuronal commands issued by the central nervous system are usually relayed faithfully by the ganglia.

Postganglionic Fibers

The postganglionic neurons that originate in the autonomic ganglia are small, unmyelinated neurons. They branch extensively and form a great variety of synaptic junctions. In some cases, the nerve terminal is enveloped in an invagination of the postsynaptic membrane, much like the synaptic cleft of skeletal muscles. In other cases, the terminal remains several microns away from its target cell.

Transmitter is released from bead-like enlargements, approximately 1 micron in diameter and 2 microns long, that appear along the nerve terminal. These varicosities contain synaptic vesicles. The vesicles in parasympathetic postganglionic fibers are similar in appearance to those found in alpha motoneuron terminals and presumably contain acetylcholine. The vesicles in the sympathetic postganglionic fibers are somewhat smaller and appear to have a dense core in electron micrographs. These vesicles are thought to contain norepinephrine.

Cholinergic Postganglionic Response

The neurotransmitter used by the parasympathetic postganglionic fibers is acetylcholine. However, the effects of acetylcholine on the postsynaptic membrane of parasympathetic postganglionic neurons are quite different from those caused by acetylcholine within the autonomic ganglia or on the skeletal muscle end plate.

It takes longer for a response to occur at the parasympathetic synapse than at the neuromuscular junction. At the end plate, a response appears within a few milliseconds after acetylcholine is released from the alpha motoneuron. By contrast, it takes approximately 100 ms before smooth muscle responds to the acetylcholine released from a parasympathetic fiber. Although some of this difference in response time may be due to the greater distance between presynaptic and postsynaptic membranes at the parasympathetic synapse, most of the difference is due to the time it takes to activate the membrane channels.

When acetylcholine combines with its receptor on smooth muscle membranes, a great variety of postsynaptic responses can occur. Some receptors are linked to channels that are selective for sodium, potassium, and calcium. Others may cause the release of calcium from extracellular and intracellular storage sites.

Chapter 5 explains that in excitation-contraction coupling of muscle this calcium can activate contraction without generating an action potential. Acetylcholine can also cause an increase in intracellular calcium, and thus in contractile activity by mobilizing membrane phosphatidylinositol or increasing intracellular cyclic-AMP levels.

The effect acetylcholine has on the heart is different from the effect that it has on skeletal or smooth muscle. In this muscle, acetylcholine causes the membrane to hyperpolarize by increasing the membrane permeability to potassium. Since the parasympathetic innervation of the heart is primarily to the pacemaker cells of the atrium, the major effect of vagal stimulation is a slowing of the heart rate.

The acetylcholine receptors on smooth muscle and the heart are pharmacologically different from those within the autonomic ganglia or on skeletal muscle. The former are classified as muscarinic and the latter as nicotinic receptors. The classification is based on the type of drugs that activate or block the two types of receptors.

Nicotinic Receptors

There are very few drugs that exclusively activate the nicotinic receptors, but as their name implies, nicotine is a fairly specific activator of the receptors in the autonomic ganglia and the end plate. Because it stimulates the ganglia, nicotine has wide-ranging systemic effects on the heart and viscera and is an extremely dangerous drug. Nicotinic receptors can be blocked by curare. Curare and its analogs have been found useful in surgical procedures where muscle relaxation is required.

Muscarinic Receptors

Muscarine is a major agonist of muscarinic receptors. Its systemic effects are similar to those of nicotine because it directly affects the parasympathetic end organs and, in addition, has a slight stimulating effect on the postganglionic cells. Atropine is the prototype of muscarinic blocking agents. It is useful as an antidote for muscarinic or anticholinesterase poisoning, for reducing the motility of the gastrointestinal tract, and for dilating the pupils in order to perform an ophthalmoscopic examination of the eyes.

Adrenergic Postganglionic Response

The postganglionic fibers of the sympathetic nervous system use norepinephrine as their neurotransmitter. The effects of the transmitter on the postsynaptic membrane vary greatly. Norepinephrine is generally inhibitory to the muscles of the gastrointestinal tract and excitatory to the vascular system and the heart. Its mode of action is not known, but appears to operate through a variety of mechanisms that include changes in membrane potential and alterations in the concentration of calcium within the postsynaptic cell.

Alpha and Beta Receptors

The sympathetic postganglionic receptors have been divided into alpha (α) and beta (β) receptors, on the basis of their responsiveness to a variety of drugs. Epinephrine is a more potent stimulator of alpha receptors than of beta receptors. Isoproterenol, on the other hand, is a much more effective stimulator of beta receptors. However, these drugs are not specific for either type of receptor. A more selective classification can be made on the basis of blocking agents. Receptors blocked by phenoxybenzamine are classified as alpha receptors, while those blocked by propranolol are classified as beta receptors.

Although very few generalizations can be made about autonomic neurotransmission, it is generally thought that activation of beta receptors causes inhibition of smooth muscle and excitation of the heart. Alpha receptors generally produce excitatory effects except within the gastrointestinal system, where they cause inhibition. Antagonists and agonists of both the alpha and beta adrenergic receptors have found wide use in medicine because of their effects on the cardiovascular, respiratory, and gastrointestinal systems.

5 Excitation-Contraction Coupling

The skeletal musculature provides the power required to maintain an upright posture and to generate movement. Following sections of the book describe the highly complex neuronal circuitry involved in the control of skeletal muscle and demonstrate that the final common pathway over which motor commands are issued is the alpha motoneuron. The synaptic process by which the alpha motoneuron generates an action potential on the skeletal muscle membrane have already been discussed. This chapter describes how this action potential causes the contraction of skeletal muscle by a process referred to as *excitation-contraction coupling*. In addition to discussing excitation-contraction coupling in skeletal muscle, this chapter also describes some of the differences that exist in the process of excitation-contraction coupling in cardiac and smooth muscle.

SKELETAL MUSCLE STRUCTURE

Skeletal muscle fibers have diameters that vary from 10 to 100 microns and may be several hundred millimeters long. The muscle fibers are organized into bundles or fascicles by a thick connective tissue sheath called the *perimysium*. Approximately 20 to 30 muscle fibers are contained in each fascicle, and individual fibers are separated from each other by another connective tissue sheath called the *endomysium*. Mechanical force developed by the muscle is transmitted from the muscle proteins through the cell membrane (sarcolemma) and connective tissue sheaths to the tendons, which are attached to the bones.

Muscle Structure

Figures 5-1 and 5-2 are diagrams of a muscle fiber, illustrating those components that are im-

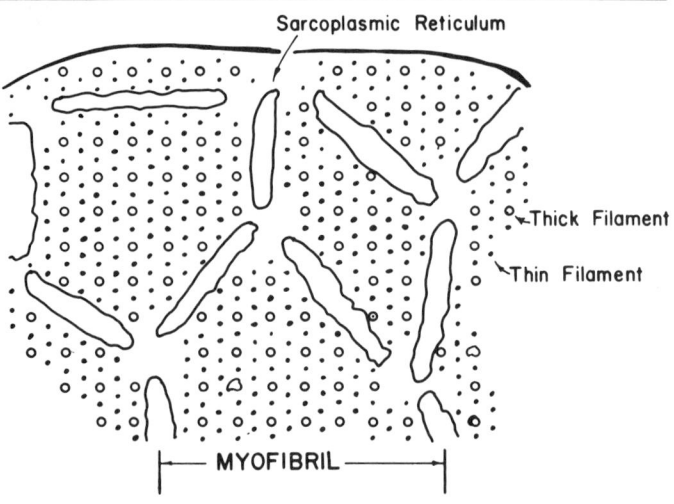

Fig. 5-1. Cross-sectional view of a muscle fiber. The sarcoplasmic reticulum divides the thick and thin myofilaments into functional units called myofibrils.

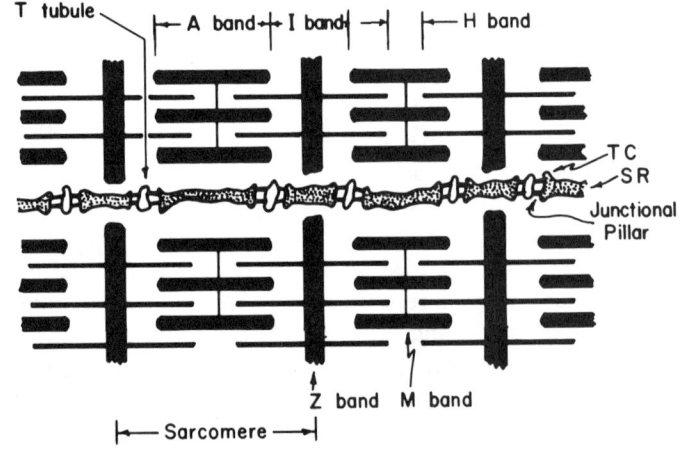

Fig. 5-2. Longitudinal view of a muscle fiber illustrating those components that participate in excitation-contraction coupling. The Z bands divide the myofibrils into functional units called sarcomeres.

portant in excitation-contraction coupling. Viewed in cross section (Fig. 5-1), the muscle fiber is seen to be divided into myofibrils, approximately 1 to 2 microns in diameter, by a system of longitudinal vesicles called the *sarcoplasmic reticulum* (SR). The myofibrils contain the thick and thin myofilaments that are responsible for generating muscular force.

The Sarcomere

Viewed longitudinally, the myofibrils are observed to be divided into functional units called *sarcomeres* by a cross-sectional band of protein called the Z *band* (Fig. 5-2). The Z bands of adjacent myofibrils are in the same plane, giving the entire muscle fiber a uniform appearance. Two types of myofilaments are contained within each myofibril, a thin filament and a thick filament.

The Myofilaments
The thin filaments are about 10 nm in diameter and 1 micron long. One end of the thin filament is attached to the Z band. The other end hangs free in the center of the sarcomere facing the thin filament, which is attached to the Z band on the other side of the sarcomere. The thick filaments (20 nm in diameter and 1.5 microns in length) interdigitate with the thin filaments in the center of the sarcomere. They are held in place by a protein bridge (the M band) that connects neighboring thick filaments to each other.

The arrangement of the thick and thin filaments within the sarcomere produces alternating dark and light bands along the muscle fiber. The A band in the center of the fiber contains the thick filaments. The I bands on either side of the A bands contain the thin filaments and are divided in two by the Z band. The portion of the sarcomere in the center of the A band, where no thin filaments overlap the thick filaments, is called the H band. When the muscle fiber is stretched or shortened, the Z bands and their attached thin filaments are pulled apart or pushed closer together. As a result, the length of the sarcomere varies. During these length changes, the overlap between the thick and thin filaments changes, resulting in a change in the length of the I and H bands.

The T Tubule and Sarcoplasmic Reticulum
In addition to the muscle proteins, Figure 5-2 also illustrates two systems of tubules within the muscle fiber that are involved in excitation-contraction coupling. The first, the T tubular network, is formed by invaginations of the muscle cell membrane. These occur at the junction between the A and I bands and branch extensively within the muscle fiber to surround the myofibrils. Its function is to conduct the action potential from the surface of the muscle fiber into its core.

The second network of tubules is the sarcoplasmic reticulum. It courses in a longitudinal direction within the spaces between the myofibrils. The SR is thin and flat throughout most of its length. However, it expands into terminal cisternae (TC) at each end where it contacts the T tubule (Fig. 5-2) at the A-I junction. The 15 nm space between the T tubule and the TC is spanned by junctional bridges or pillars, which presumably are the site of communication between the T tubule and the SR. The SR contains a high concentration of calcium and is responsible for releasing calcium into the cytoplasm during the process of excitation-contraction coupling.

ACTIVATION OF SKELETAL MUSCLE
The Thick and Thin Filaments
The generation of force by muscle fibers is carried out by the proteins contained in the thick and thin filaments. The thick filament is composed of about 300 myosin molecules that are approximately 2 nm in diameter and 160 nm long. The myosin molecules are divided into two sections; one, called *light meromyosin*, is embedded in the thick filament and the other, called *heavy meromyosin*, extends toward the thin filament (Fig. 5-3).

The free end of the heavy meromyosin expands to form a globular head about 50 nm long called the *cross bridge*. Although not shown in Fig. 5-3, the end of each myosin molecule contains two globular heads and a series of light protein chains. These light chains play an important role in excitation-contraction coupling. For example, the myosin ATPase, used to provide energy for muscle contraction, is contained within one of the light chains.

The thin filaments are composed of three proteins (Fig. 5-3). The backbone of the thin filament is formed by two intertwined chains of the protein actin. During contraction, the myosin

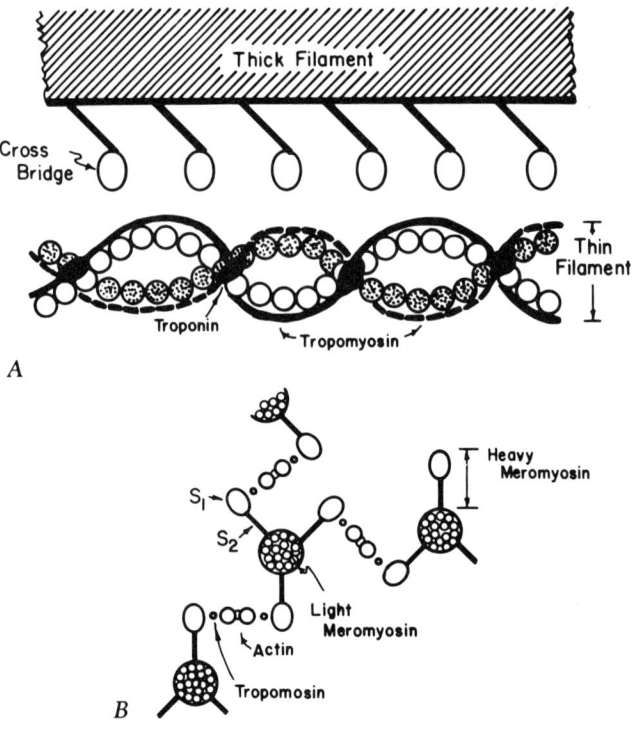

Fig. 5-3. Longitudinal view of the thick and thin filaments (A). The heavy meromyosin molecules with their cross bridges are extended from the thick filament toward the thin filament (B). The tropomyosin is positioned between the thick and thin filaments and thus prevents interaction between actin and myosin.

cross bridge attaches to a binding site on the actin molecule. When the muscle is at rest the cross bridge cannot attach to the actin, because the latter is covered by another of the thin filament proteins, tropomyosin. A third thin filament protein, troponin, regulates the position of the tropomyosin.

Excitation-Contraction Coupling

The first step in skeletal muscle excitation-contraction coupling is the generation of an action potential on the skeletal muscle fiber membrane. The action potential spreads over the surface of the membrane and into the T tubular system, where it causes the release of calcium from the SR. The calcium released from the SR binds to troponin and causes it to undergo a conformational change. The alteration in the troponin causes tropomyosin to move away from its position between the myosin and actin molecules. Once this occurs, the cross bridge binds spontaneously to actin and contraction begins.

Release of Calcium from the SR (The Active State)

Recall from the discussion in Chapter 4 that each time an action potential propagates into the terminal of an alpha motoneuron, enough transmitter is released to cause the initiation of an action potential on the skeletal muscle membrane. This action potential is then propagated along the muscle and T tubular membranes. Calcium is released from the terminal cisternae of the SR when the T tubular membrane attached to it is depolarized. However, the mechanism by which calcium is released is not understood.

Although the junction between the T tubule and the SR has some similarities to the gap junction discussed in Chapter 4, there are no low resistance electrical pathways over which the current generated by the T tubular action potential could depolarize the SR. Several theories have been proposed to explain how the depolarization of the T tubule might cause the release of calcium from the SR. One theory suggests that the electric field created by the T tubular action potential opens a gate in the SR, through which calcium can flow into the cytoplasm of the muscle fiber. Alternatively, channels in the T tubular membrane could open during the action potential, allowing a small amount of calcium to enter the cell. Although the amount of calcium entering the cell through these channels would be too small to cause a conformational change in troponin, it could cause the release of calcium from the SR. This mechanism is referred to as the *trigger calcium* or *calcium-induced calcium release hypothesis*. Regardless of which of these theories is correct, if any, the result of a T tubular action potential is the release of enough calcium from the SR to raise the intracellular calcium concentration from its resting value of 10^{-7} mol/L by 100-fold, to 10^{-5} mol/L. This concentration of calcium is sufficient to bind to all of the troponin molecules within the skeletal muscle fiber and thus activate all of the muscle proteins. All of the troponin molecules thus undergo a conformational change, so that tropomyosin moves away from all of the myosin binding sites on the thin filaments. Once the binding sites are free of inhibition, the globular head of the myosin cross bridge binds to the thin filament and initiates the process of contraction. The period of time during which the myoplasmic calcium concentration is high enough to cause activation of the muscle proteins is called the *active state*.

The Cross-Bridge Cycle

The second step in excitation-contraction coupling is the cross-bridge cycle that begins when myosin binds to the actin molecules on the thin filament. The cross-bridge cycle (Fig. 5-4) consists of the binding, bending, and detachment of the cross bridge.

Each cross-bridge cycle moves the thin filament by less than 50 nm. However, cross-bridge cycling continues for as long as the calcium concentration remains above 10^{-7} mol/L. As a result of this repeated cross-bridge cycling, a significant amount of shortening and force development can take place.

The energy required for the cross-bridge cycle is provided by ATP. As indicated above, the ATPase required to liberate the energy of the ATP is contained in one of the myosin light chains. Also, as illustrated in Figure 5-4, the ATP utilized by the cross bridge is bound to the globu-

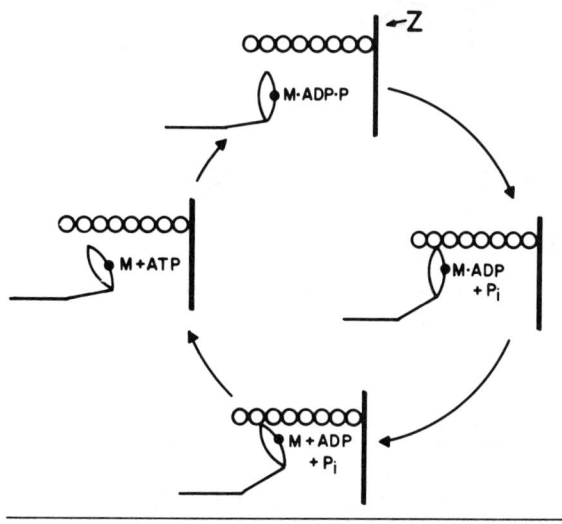

Fig. 5-4. Steps of the cross-bridge cycle. Before contraction, the cross bridge is at right angles to the thick filament. When it binds to actin, the high energy myosin (M)-ADP-P_i complex is hydrolyzed. The energy obtained from this reaction is used to bend the cross bridge. The products of hydrolysis are then released from the cross bridge and a new molecule of ATP is added causing the cross bridge to detach from actin. The cross bridge then returns to its original position and the cycle is repeated.

lar head of myosin. Although the exact sequence of steps involved in the transfer of the energy contained in ATP to the mechanical work of the cross-bridge cycle is not completely understood, Figure 5-4 indicates one way in which it might occur.

At rest, the ATP molecule is partially hydrolized by the ATPase to the high energy intermediate form, ADP-P_i. When the cross bridge combines with actin, the ATPase becomes fully activated and completes the hydrolysis, converting ADP-P_i to ADP and P_i. During this step, the energy of ATP is used to bend the cross bridge.

The cross bridge cannot detach from the thin filament until a new molecule of ATP binds to it. When this occurs, the cross bridge returns to its upright position, ready to begin another cycle. If ATP is not available, cross-bridge detachment cannot occur. This is the reason for the muscle stiffness associated with rigor mortis.

The Sliding Filament Hypothesis

The sequence of events described above to explain muscle contraction is called the *sliding filament theory*, because the repetitive cycling of cross bridges causes the thick and thin filaments to slide across each other. One of the strongest supports for this hypothesis is the relationship between muscle length and force development (the length-tension relationship).

The Length-Tension Relationship

If the generation of force is due to the attachment of cross bridges to the thin filaments, then the amount of force developed during a contraction should be proportional to the number of myosin cross bridges that bind to actin. This number can be varied by varying the muscle length. If a resting muscle is stretched, the amount of overlap between the thick and thin filaments, and consequently the number of cross-bridge attachments that can form, vary. Figure 5-5 illustrates the result of an experiment in which the amount of force generated by a muscle at different muscle lengths was measured.

When the muscle is stretched so that its sarcomeres are 3.65 microns long, no force is developed by the muscle when it is stimulated. At this muscle length, the thin filaments are pulled so

Fig. 5-5. Force generated by a muscle at various sarcomere lengths.

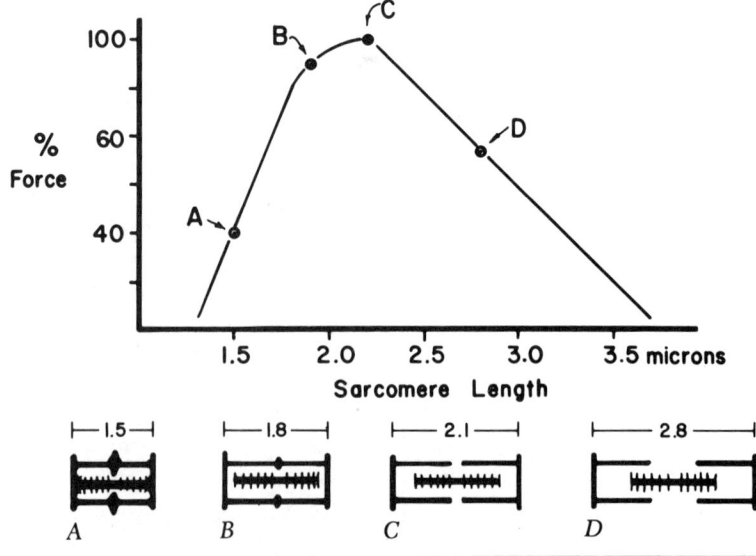

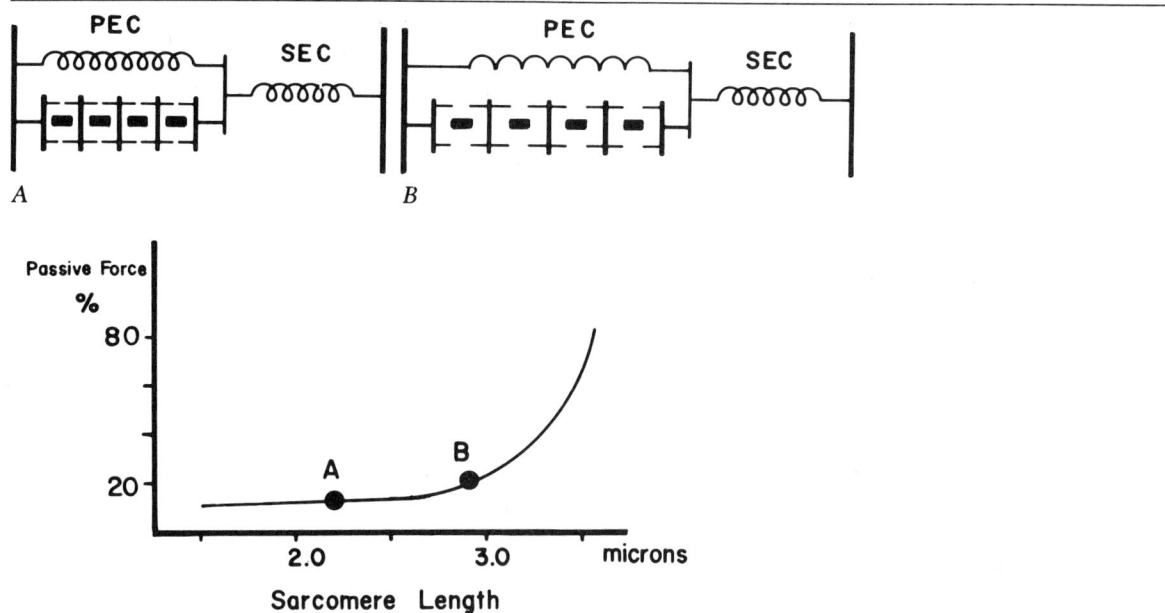

Fig. 5-6. The parallel elastic component (PEC) causes the muscle to resist being stretched. The PEC is fairly compliant so that little resistance occurs until the sarcomere is stretched to 3.0 microns.

far from the center of the fiber that the thick and thin filaments do not overlap at any point within the sarcomere. So although calcium is released from the SR and tropomyosin no longer inhibits the binding of actin and myosin, no force is developed because there are no actin molecules opposite any of the cross bridges.

If the initial sarcomere length is shortened to 2.20 microns, a maximal amount of force is developed by the muscle. This is because all of the cross bridges on the thick filament are able to bind to an actin molecule on the thin filament during contraction. As the muscle is stretched from an initial length of 2.20 microns to 3.65 microns, the number of cross bridges that can bind to actin during a contraction decreases and as a result, the force developed by the muscle decreases.

At sarcomere lengths between 2.2 and 2.0 microns, the force developed by the muscle remains at its maximum level, because there is no change in the number of cross bridges that can form. At sarcomere lengths below 2.0 microns, the force developed by the muscle declines. This decline most likely results from the mechanical impediments to contraction that occur first as the thin filaments meet in the center of the sarcomere (2.00 to 1.65 microns), and second as the Z band pushes up against the thick filament (1.65 to 1.35 microns).

Muscle Relaxation

The final step in excitation-contraction coupling is muscle relaxation. This occurs when the cytoplasmic calcium concentration is reduced to 10^{-7} mol/L by the SR. The flat, longitudinal portion of the SR contains a high concentration of calcium pumps. These are ATPase molecules that actively transport calcium in a manner analogous to the sodium-potassium pump described in the first chapter. They are able to rapidly sequester calcium into the SR. The amount of calcium that can be stored within the SR is increased by the calcium binding protein calsequestrin.

Although the amount of force developed by the contractile proteins depends on the number of cross bridges formed, the actual amount of force transmitted to the tendons and bones is determined by the structural elements connecting the muscle proteins to the bones. Figure 5-6 is a diagram depicting these structural elements. There are two elastic elements that need to be considered, the parallel elastic component (PEC) and the series elastic component (SEC).

The Parallel Elastic Component

The parallel elastic component (PEC) is, as its name implies, in parallel with the contractile elements of the muscle fiber. When the muscle fiber is stretched, the resistance to the stretch is provided by the PEC. As shown in Figure 5-6, very little force is required to passively stretch the muscle fiber sarcomere from a resting value of 2.0 microns to approximately 2.8 microns. Above 3.0 microns, however, the resistance of the PEC to passive stretch increases markedly.

The high compliance of the PEC in skeletal muscle to moderate amounts of stretch allows the muscle length to be varied over its operational length. As the muscle length approaches 3.6 microns (the length at which no contractile force can be developed), the PEC becomes extremely stiff and effectively prevents the muscle from being stretched any further. The sarcomere itself does not offer any resistance to stretch.

The Series Elastic Component

The other elastic element that affects the contractile behavior of the muscle is the series elastic component (SEC). This elastic component is placed between the contractile proteins and bone so that any force developed by the muscle must be transmitted through it. The SEC is much stiffer than the PEC, so that it does not change length during passive stretch of the whole muscle. However, as shown in Figure 5-7, when the muscle contracts the SEC lengthens in proportion to the amount of force developed. Because the SEC is stretched during a contraction, the amount of

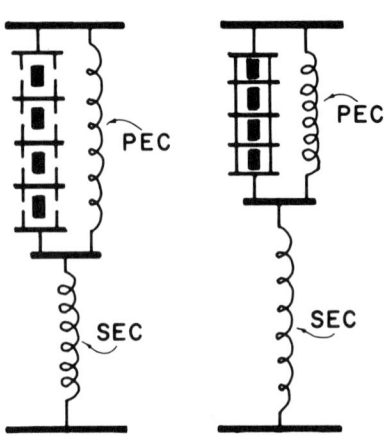

Fig. 5-7. *When a muscle contracts, the series elastic component (SEC) is stretched. The greater the amount of force developed, the more the SEC is stretched.*

force developed by the contractile proteins is not immediately transmitted to the bones.

Each cross-bridge cycle brings the Z bands closer together and stretches the SEC. The more the cross bridges cycle, the more the SEC is stretched, and the greater the amount of force transmitted to the bone becomes. Since the cross bridges continue to cycle as long as the calcium concentration remains above 10^{-7} mol/L, the force transmitted to the bone depends on the duration of the active state.

Summation and Tetanus

The duration of the active state in the muscle, illustrated in Figure 5-8, is approximately 150 ms. That is, it takes 150 ms for all of the calcium released from the SR to be resequestered. During this time there are only enough cross-bridge cycles to stretch the SEC to one third of its maximal length. Thus, during a single muscle twitch, the force transmitted to the bone is only one third of the force developed by the muscle proteins.

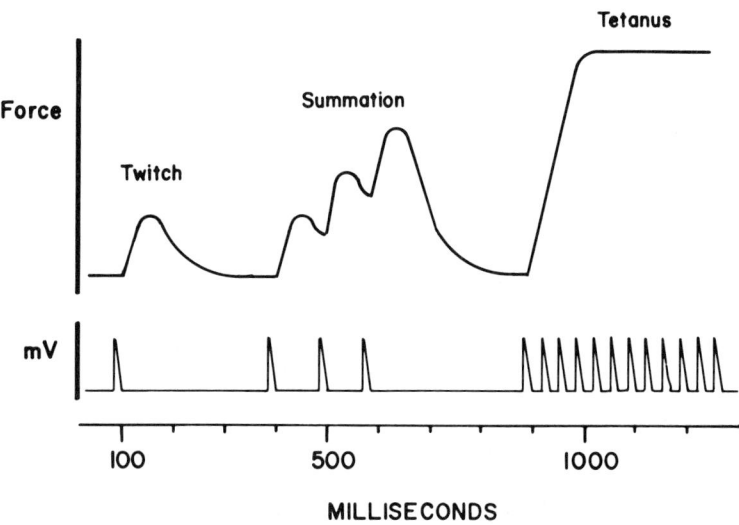

Fig. 5-8. The amount of force developed by a skeletal muscle can be increased by increasing the frequency of action potential discharge. Force is increased because calcium remains in the myoplasm for a longer time when the action potential frequency is increased; thus, more cross-bridge cycles can occur.

The amount of force transmitted to the bone can be increased, as shown in Figure 5-8, by increasing the frequency of muscle contraction. Because the active-state duration is much longer than the refractory period of the muscle action potential, a second action potential can be generated before all of the calcium released by the first action potential is resequestered by the SR. By increasing the frequency of stimulation, the calcium remains within the cell for a longer period of time. This keeps the cross bridges cycling longer and, as a result, the force transmitted to the bone increases.

As illustrated in Figure 5-8, the contractile force produced by each subsequent action potential adds to the force produced by the previous action potential. If the stimulation frequency is made high enough, then the force does not diminish between action potentials and all the force generated by the muscle proteins is transmitted to the bone. This smooth rise to maximum tension is called *tetanus*.

The principles of excitation-contraction coupling described here for skeletal muscle also apply to cardiac and smooth muscle. However, there are some important differences.

ACTIVATION OF CARDIAC AND SMOOTH MUSCLE

Cardiac muscle, like skeletal muscle, is organized into sarcomeres. It possesses a T tubular system to propagate an action potential into the core of the muscle fiber and an SR to release and sequester calcium. In cardiac muscle, the T tubules are located at the Z band rather than at the A-I junction, as they are in skeletal muscle. Also, some of the terminal cisternae of the SR form junctions with the muscle membrane in addition to the junctions made with the T tubule. Because cardiac muscle fibers are much smaller (about 10 to 20 microns wide and 5 microns thick) than skeletal muscle fibers, they are less dependent on the T tubular system for excitation-contraction coupling.

Excitation-Contraction Coupling in Cardiac Muscle

One important difference between skeletal and cardiac muscle is that the calcium used for excitation-contraction coupling in cardiac muscle does not come exclusively from the SR. Instead, a significant portion of it probably comes from outside the cell. Figure 5-9 illustrates a cardiac action potential and a cardiac muscle twitch.

Notice that unlike nerve and skeletal muscle action potentials, the cardiac action potential has a long plateau. Although a description of the cardiac action potential is beyond the scope of this book, it should be noted that in addition to containing sodium and potassium channels similar to those described for nerve and skeletal muscle, cardiac muscle also contains calcium channels. These are open during the plateau, and the calcium flowing through them participates in muscle contraction.

Another important difference between cardiac and skeletal muscle is that the concentration of calcium entering the cytoplasm during the active state is never sufficient to activate all of the troponin. As a result, the number of cross bridges formed during a twitch is less than would be predicted by the amount of overlap between the thick and thin filaments. Also, the parallel elastic component in cardiac muscle is much stiffer than it is in skeletal muscle. Consequently, the initial length of cardiac muscle sarcomeres is never greater than 2.2 microns and is usually much less. Both of these factors, the amount of calcium entering the cytoplasm and the initial length of the cardiac sarcomere, are under physiological control and are varied to vary the amount of force generated by the heart.

A final difference between cardiac and skeletal muscle is that in cardiac muscle, the durations of the action potential and muscle twitch are approximately the same (see Fig. 5-9). As a result, the duration of the active state can not be prolonged by repetitive stimulation, and summation in cardiac muscle cannot occur.

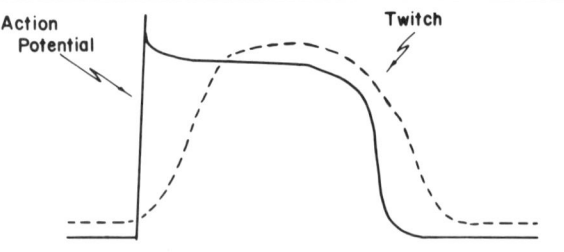

Fig. 5-9. *Cardiac action potential and the muscle twitch produced by it.*

Excitation-Contraction Coupling in Smooth Muscle

Smooth muscle is structurally quite different from skeletal and cardiac muscle. It is not organized into sarcomeres and does not contain T tubules. And although it possesses a sarcoplasmic reticulum, most of the calcium used for excitation-contraction coupling in smooth muscle fibers probably comes from the extracellular fluid. A myriad of methods, in addition to the release of calcium from the SR by an action potential, are utilized by smooth muscles to increase their intracellular calcium concentrations.

Some smooth muscles possess calcium channels through which calcium flows into the cell during the action potential. In other smooth muscles, calcium enters the cell through channels that are opened by neurotransmitters or by membrane depolarization. In still others, calcium is released from intracellular storage sites, without any change in membrane potential, by a mechanism referred to as *pharmacomechanical coupling*. Despite these differences, the sliding of thick and thin filaments across each other is the basic mechanism of contraction and is the same as in skeletal and cardiac muscle.

The role of calcium in excitation-contraction coupling in smooth muscle, however, differs dramatically from that occurring in skeletal or cardiac muscle. In skeletal and cardiac muscle, cross-bridge cycling is inhibited in the resting state by tropomyosin. When this inhibition is re-

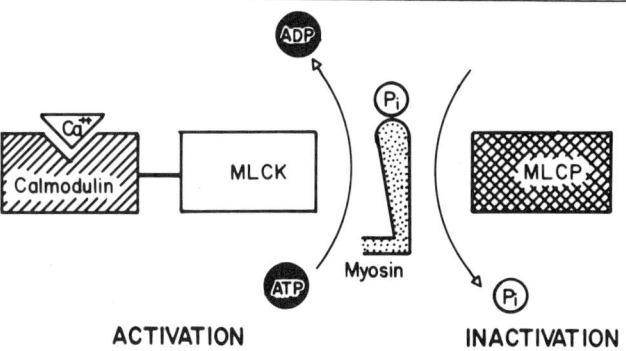

Fig. 5-10. Excitation-contraction coupling in smooth muscle requires phosphorylation of the myosin light chains. Phosphorylation occurs when an enzyme, myosin light chain kinase (MLCK), is activated by calcium. Dephosphorylation is brought about by the enzyme, myosin light chain phosphatase (MLCP).

moved by the calcium-induced conformational change in troponin, myosin spontaneously combines with actin. In smooth muscle, there are no thin filament regulatory proteins preventing the interaction of myosin and actin. Instead, cross-bridge interactions do not occur in the resting state, because the myosin is in an inactive form.

Activation of myosin involves a series of reactions culminating in the phosphorylation of one of the myosin light chains. The sequence is begun by the binding of calcium to a cytoplasmic protein called calmodulin (Fig. 5-10). Calmodulin then binds to an enzyme, myosin light chain kinase (MLCK), which catalyzes the phosphorylation of the cross bridge. One ATP molecule is consumed for each myosin light chain that is phosphorylated. Once in the phosphorylated, active form, myosin binds to actin and the cross-bridge cycle is initiated.

Cross-bridge cycling continues until the myosin light chain is dephosphorylated. This step is controlled by the enzyme, myosin light chain phosphatase. Although this reaction occurs throughout the active state, dephosphorylation is a much slower process than phosphorylation, and thus not too many ATP molecules are required to maintain myosin in its active form. However, when calcium is removed from the cytoplasm, the phosphatase enzyme dephosphorylates the myosin light chain, and cross-bridge cycling comes to an end. The number of reactions involved in excitation-contraction coupling has made the regulation of smooth muscle contractile activity difficult to understand. Much more work has to be done in this area before its true nature is revealed.

II Sensory Physiology

6 Receptors

Our perception of the world around us is achieved through the action of our sensory receptors. These receptors, in the skin, eyes, ears, tongue, and nose, receive environmental stimuli and transform them into neural activity that is then transmitted to the central nervous system. The sensory information is used to initiate reflexes, to aid in the production of movement, and to provide us with conscious sensations. This chapter explains the general mechanisms by which all sensory receptors operate. Subsequent chapters are devoted to detailed descriptions of the specific receptors responsible for our sensory perceptions.

CLASSIFICATION OF RECEPTORS

Receptors can be classified on the basis of the type of sensation they produce, their location within the body, their physiological function, or the type of energy that activates them.

Exteroceptors

Receptors that detect stimuli originating outside of the body are called *exteroceptors*. These receptors are commonly classified by the type of perception they produce. Using this scheme, the five primary sensory modalities can be described as sight, sound, touch, taste, and smell. Each of these categories is further divided into submodalities, such as red and blue for vision or sweet and sour for taste.

Interoceptors

Another group of receptors, called *interoceptors*, monitor the internal condition of our physiological state. Unlike the exteroceptors, the interoceptors do not produce a specific conscious sensation of the stimuli they detect. Thus, we do not know our blood pressure or the concentration of glucose in our blood despite the presence of receptors responsible for monitoring their values. However, the activity of these receptors may produce a general awareness of our physiological condition. For example, the detection of low blood sugar levels may lead to the sensation of hunger. Or an increase in blood osmolarity may lead to the perception of thirst.

Proprioceptors

Still another group of receptors, called *proprioceptors*, is responsible for detecting the position of our bodies in space and the activity of our skeletal muscular system. Like the interoceptors, they create a general awareness of body position and movement without producing specific sensations. For example, we do not know the exact length of our muscles or the tension that they are developing. Yet our muscle activities are accompanied by a perception of effort that is related to the activity of our proprioceptors. Similarly, no specific sensation results from the stimulation of the vestibular apparatus, but because of its activity we are provided with an awareness of "up" and "down."

The simplest and most inclusive way to classify receptors is based on the type of stimulus energy that activates them. Using this scheme, receptors can be divided into mechanical (which include, among others, both the receptors for the conscious sensations of touch and hearing and those for the unconscious detection of blood pressure), thermal, chemical (which include taste and glucose receptors), and photoreceptors.

Regardless of how they are classified, the function of all sensory receptors is to convert an environmental stimulus into a neuronal signal that can be understood by the central nervous system. This chapter considers those functions that are common to all receptors and then, in subsequent chapters, deals with the operation of specific receptors.

ACTIVATION OF RECEPTORS

The detection and subsequent transmission of sensory information to the central nervous system is a two-step process. The receptor first transforms the stimulus into a local change in membrane potential, called the *receptor or generator potential*. Then, the generator potential serves as a stimulus for the generation of action potentials that propagate along the sensory axon into the spinal cord or brain.

The Transducer

The membrane component responsible for converting the sensory stimulus into a receptor potential is called the *transducer membrane* (Fig. 6-1). This membrane is similar to the postsynaptic membrane in that it does not contain the time- and voltage-dependent gates that are necessary to produce an action potential. Thus, the transducer membrane does not produce an all-or-none response to a stimulus. Instead, like the postsynaptic membrane potential, the magnitude of the receptor potential is proportional to the intensity of the sensory stimulus.

The stimulus that normally activates a receptor is called an "adequate stimulus" to indicate that extremely small amounts of stimulus energy are able to produce a response. Other stimuli are capable of activating the receptor, but to do so they require much greater stimulus intensities. For example, although a single photon of light (the adequate stimulus for vision) is sufficient to elicit a visual response, a mechanical blow to the eyes also evokes a visual sensation.

The mechanical receptor responsible for the sensation of touch is often used to describe the general mechanisms of receptor function. The transducer membrane of this receptor is located at the end of a sensory neuron. When the transducer is deformed by a mechanical stimulus, the membrane conductance for sodium and potassium increases, resulting in a depolarizing receptor potential. However, all receptors do not act in this fashion. The mechanism by which the receptor potential is produced varies from receptor to receptor. In addition, the transducer region is not always found on the nerve terminal of the sensory neuron nor is the receptor potential always a depolarization of the transducer membrane.

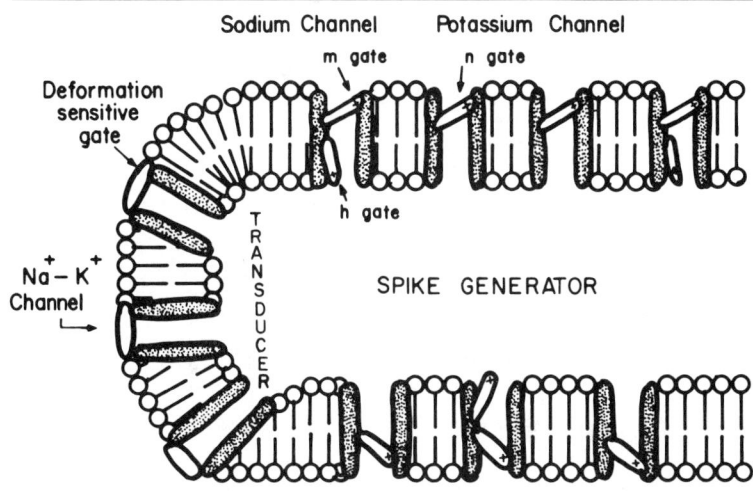

Fig. 6-1. *Components of a receptor.*

Chemoreceptors, for example, are activated when the chemical acting as an adequate stimulus for the receptor combines with a binding site on the transducer. The stimulus-receptor complex causes the opening of ionic channels within the transducer. Ionic current then flows through these channels to produce the receptor potential. This process is similar to the mechanism described earlier by which synaptic transmitters activate postsynaptic membranes.

Although the transducer for the olfactory chemoreceptor is found on the olfactory nerve terminals, the transducer for the taste chemoreceptor is located on a modified epithelial cell within the tongue. The receptor cell, as described in the chapter on taste, releases a synaptic transmitter that stimulates the cranial nerves for taste sensation.

In the eye, light causes a photochemical reaction in chromophores contained within the photoreceptors. The membrane potential of the photoreceptor cell is changed as a consequence of the photochemical response. However, the transducer membrane does not depolarize. Instead, as described in Chapter 9 the receptor potential in the eye is a hyperpolarizing potential.

How receptor potentials are produced in thermoreceptors is not known. However, it is possible that the transducer membrane potential of these receptors is maintained by an electrogenic metabolic pump that is highly sensitive to temperature changes. If so, then small changes in temperature could cause the membrane potential changes observed when these receptors are exposed to changes in temperature.

Regardless of the receptor type or the mechanism by which the receptor potential is produced, the only way for information to be transmitted from the receptor to the central nervous system is for an action potential to be elicited and propagated along the sensory axon. This task is accomplished by the spike generating region of the receptor.

The Spike Generator

The spike generating region or trigger zone of the receptor is located next to the transducer (Fig. 6-1). The membrane of the spike generator contains the time- and voltage-dependent gates required for the production of action potentials. Thus, when the depolarization of the transducer membrane depolarizes the spike generating membrane to threshold, an action potential is generated.

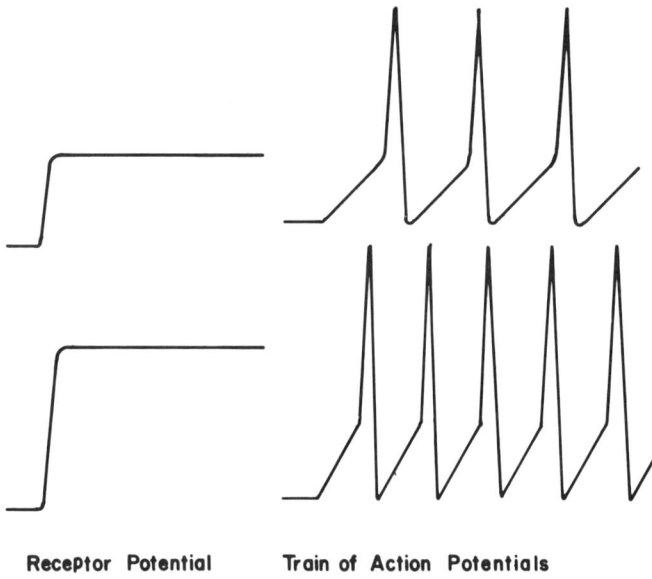

Fig. 6-2. *At the conclusion of each action potential, the receptor potential causes the membrane to depolarize toward threshold.*

As illustrated in Figure 6-2, the receptor potential produces repetitive firing of action potentials by the spike generator. This occurs because, after the completion of one action potential, the receptor potential causes the spike generating region of the receptor to be depolarized to threshold once again. The rate at which threshold is reached depends on the magnitude of the receptor potential. A greater receptor potential causes threshold to be reached sooner and consequently causes a greater frequency of action potential discharge.

Adaptation

Not all receptors produce a steady stream of action potentials in response to the steady application of a stimulus. In the majority of receptors, some degree of adaptation takes place; the frequency of action potentials decreases even though the stimulus is maintained. Those receptors that continue to generate action potentials for as long as the stimulus is present are called *nonadapting*, *static*, or *tonic receptors*. Those that reduce their rate of firing are called *adapting*, *dynamic*, or *phasic receptors*.

There are two main mechanisms by which adaptation takes place. In one, adaptation is caused by an alteration in the mechanical linkage between the stimulus and the transducer portion of the sensory axon. For example, in the pacinian corpuscle (examined in the Chapter 7, Cutaneous Sensation), there is a complex capsule surrounding the nerve terminal that transmits mechanical deformation of the skin to the sensory transducer (Fig. 6-3). The capsule is constructed so that only rapidly changing deformations are transmitted to the nerve terminal. As soon as the deformation reaches a steady state, the layers within the capsule become reorganized so that the deformation at the periphery of the capsule no longer reaches the nerve terminal. Consequently, the magnitude of the receptor potential decreases and the action potential firing frequency diminishes.

The other mechanism responsible for pro-

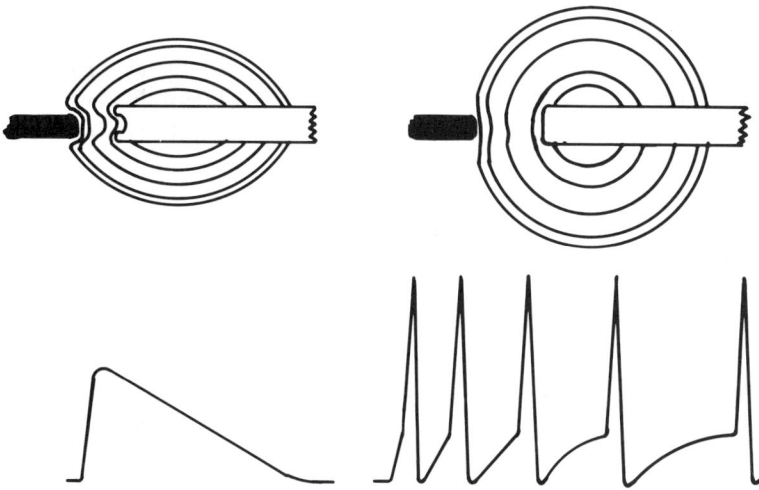

Fig. 6-3. In the pacinian corpuscle, adaptation occurs because only rapidly changing stimuli are transmitted to the transducer region of the sensory axon. If the stimulus is applied for more than one second, the connective tissue layers of the capsule become rearranged and the transducer is no longer deformed.

ducing adaptation is based on the characteristics of the membrane in the vicinity of the spike generating region. Although the experimental evidence for this is not conclusive, it appears that the membrane surrounding the spike generating region increases its permeability to potassium when depolarized. The increase in potassium permeability causes the spike generating region to repolarize despite the presence of the receptor potential. As a result, the frequency of firing is reduced.

SENSORY CODING

Information about the nature of the stimulus is encoded as a train of action potentials by the primary sensory axons. It is then transmitted into the central nervous system and ultimately to the cortex, where it results in a conscious perception of the stimulus. The actual mechanism by which the firing of action potentials in sensory neurons gives rise to consciousness is not at all understood. However, certain ideas about how information is encoded by the nervous system are sufficiently well established to warrant discussion.

Encoding of Stimulus Intensity

The first step in the process of sensory coding is the transduction of stimulus energy into a receptor potential. The relationship between the magnitude of the stimulus and the size of the generator potential can be expressed as either a logarithmic function, Receptor Potential = K (log stimulus), or a power function, Receptor Potential = K (stimulus)^n, where K and n = constants.

The logarithmic relationship is referred to as the Weber-Fechner law after the scientists who first described it. The power relationship is called the Stevens power law. Both laws indicate that as the magnitude of the stimulus rises, there is a corresponding yet smaller increase in the size of the receptor potential. This is of functional importance, because it allows a sensory system to respond over a wide range of stimulus energies without saturating.

The maximum amount of depolarization caused by a receptor potential is approximately

50 mV. If a minimally detectable stimulus requires a potential change of 1 mV, then the range of stimulus intensities would be limited to approximately 50 to 1. However, because of the relationship between stimulus magnitude and receptor potential, indicated by the equations given previously, the range of stimulus intensities can be greatly extended. For example, suppose the equation relating the stimulus intensity and receptor potential for a mechanical receptor is Receptor Potential = (stimulus)$^{0.5}$. In this case, the stimulus intensity could increase to 2,500 times threshold before the receptor potential reaches 50 mV and the receptor saturates. Even greater ranges of stimulus intensities can be accommodated by receptors in which the exponent in the above equation is less than 0.5. This relationship, while allowing for an increase in the range of stimulus intensities that can be encoded, also causes a decrease in our ability to detect a small difference in stimulus intensities as the magnitude of the stimulus increases. Compare, for example, the smallest difference in stimulus magnitude that can be detected when the stimulus is at threshold to the difference in magnitude required for detection when the stimulus is 1000 times threshold.

Using the assumption that a 1 mV change in receptor potential is required for detecting a difference in stimulus intensity, increasing the stimulus strength from a value of 1 to 4 units is sufficient for detection. However, if the stimulus strength is 1000 units, then its value must be increased to 1064 in order to increase the receptor potential by one mV. This trade-off between expanding the range of stimulus magnitudes that can be detected and the loss of sensitivity at high intensities is typical of the compromises made by the nervous system to use its inherent capabilities and limitations most efficiently to meet conflicting needs.

The relationship between the frequency of action potential firing and stimulus intensity is also described by the Weber-Fechner and Stevens laws. Thus, it appears that the stimulus intensity is encoded by the frequency of action potentials generated in the sensory neuron. The validity of this conclusion is strengthened by experiments showing that our perception of stimulus intensity is also related to stimulus strength by the same Weber-Fechner and Stevens equations that describe the relationship between firing frequency and stimulus strength.

Encoding the Rate of Stimulus Application

As indicated earlier, the frequency of action potential firing encodes the intensity of the applied stimulus. If the firing of action potentials were to slow down or cease, then the information about stimulus intensity would be lost. Thus, the encoding of intensity requires the use of a nonadapting or tonic type of receptor. Phasic receptors, on the other hand, encode the rate of stimulus application.

Phasic receptors adapt when a stimulus of constant intensity is applied to them. In order to maintain activity in a phasic receptor, the intensity of the stimulus must be continuously increasing. By applying the stimulus rapidly, there is little opportunity for adaptation to take place and a high rate of action potential firing is obtained. If, however, the stimulus is applied slowly, the amount of adaptation is significant, and the firing frequency is low. Thus, the frequency of firing in a phasic receptor is proportional to the velocity of stimulus application.

The use of adaptation as a mechanism of sensory encoding underscores the concept that perceptual adaptation is not caused by sensory adaptation. Perceptual adaptation occurs whenever a sensory stimulus is applied for a long time. For example, we are usually not aware of the pressure exerted by the chair on our buttocks nor of the humming of an electric fan. Yet these stimuli continue to produce tonic activity in the sensory neurons responsible for encoding their presence.

Perceptual adaptation is thus not a function of sensory adaptation but rather a function of centers within the brain. Almost everyone has had the experience of being in a room where multiple conversations are occurring and being able to concentrate on a single conversation. However, if your name is mentioned in one of the other conversations, your attention is immediately drawn to it and, most remarkably, you discover that you are aware of the context in which your name was mentioned. In other words, even though you were not conscious of information received from the other conversation, your brain obviously received it. If your attention were not drawn to the conversation, you would never have become aware of what was being said. More scientific studies have been done to demonstrate that not all of the information detected by your receptors and received by your brain results in a conscious perception.

The ability to direct nonessential information away from the centers of consciousness prevents the brain from being overwhelmed by the multitude of stimuli that are continuously impinging on it. At the same time, it allows the information to be received, decoded, and if important, to become the focus of our attention. It is likely that these centers do not function properly in hyperactive children, so that they are unable to maintain attention on a single task. Instead, their focus continuously shifts as each new stimulus becomes the object of their attention.

Encoding of Sensory Quality
Although the mechanisms by which stimulus intensity and velocity are encoded by the nervous system seem fairly well understood, there is less known about how the quality of a stimulus is transmitted to the brain.

The simplest way to encode quality is through a *labeled line*. This mechanism requires that a neuronal pathway convey information about only one type of sensory stimulus. The brain is able to decode the sensory quality of the stimulus because of where the pathway terminates within the cortex. For example, activation of neurons within the somatosensory cortex yields a sensation of touch that is projected to the area of skin from which the labeled line neurons originated. Similarly, cortical cells within the auditory area of the temporal lobe and the visual areas of the occipital lobe produce sensations of sound and light, respectively, when they are stimulated.

Each of the primary sensory modalities is encoded by a labeled line. Thus, no matter what mechanism is used to stimulate the receptor, the sensation perceived is dependent on the neural pathway activated and not on the stimulus used. For example, whether the retina is stimulated by light or a mechanical blow to the head, the sensation is always visual, never mechanical. Although labeled lines are a precise way of encoding sensory stimuli, they are not very efficient, because each quality of sensation requires its own labeled line.

A more efficient encoding mechanism can be devised if each neuron is able to participate in the coding of several different sensory qualities. Using such a system, three neurons can encode seven qualities if each of the possible combinations of neuronal firing were capable of encoding a different stimulus quality. For example, the firing of neurons No. 1 and No. 2 could signal one stimulus, the firing of neurons No. 2 and No. 3 could indicate another stimulus, and the firing of all three neurons together could encode a third stimulus.

Even more sensory qualities could be encoded by these neurons if their actual firing pattern were to indicate a different sensory quality, for example, if a neuron were to generate a steady firing rate in response to one stimulus and a bursting firing pattern in response to another.

Feature Detectors. Another way in which quality can be encoded is through using *feature detectors*. These are neurons within the brain that respond only when a particular group of sensory neurons are activated. For example, specialized neurons within the auditory system, that respond only when activated by sensory input from both ears, are used to localize a sound in space. Other feature detectors in the visual system are used to detect the contrast, contour, and movement of a visual signal. These and other feature detectors are discussed in the chapters covering their specific receptors.

A great deal more needs to be learned about how sensory information is encoded by the nervous system. The success of current efforts to develop prosthetic devices that can be used to replace damaged eyes and ears depends on gaining a thorough understanding of sensory coding.

7 Cutaneous Sensation

The skin is an extremely important organ that serves a variety of functions. It forms a waterproof shield around the internal organs that prevents excessive water loss and protects against the invasion of harmful microorganisms. In addition, it plays an essential role in temperature regulation and in the formation of vitamin D. Finally, it receives a rich neural innervation, permitting us to become aware of that part of the external environment that comes into direct contact with our skin.

The skin is divided into hairy and hairless (glabrous) skin. All of the body, except for parts of the face and the palmar surface of the hands and feet, is covered with hairy skin. Apart from the obvious difference of not having any hair, glabrous skin contains many more sweat glands, has a thicker epidermal covering, and has a greater density of sensory nerve endings. Since the specialized sensory endings associated with hairs do not play a significant role in the sensory perceptions of humans, no distinction between the sensory capabilities of the two types of skin is made here and only those receptors that appear in both areas are described.

The sensory perceptions generated by the activity of receptors found within the skin can be classified as touch, pain, and temperature. The sense of touch is usually extended to include the perception of pressure and vibration also. This chapter discusses those receptors involved in the production of our perception of touch, pain, and temperature.

THE TOUCH RECEPTORS

Over 1,000,000 sensory axons innervate the skin, terminating for the most part within the dermis (just beneath the epidermis). The great majority of these fibers are distributed to the face and hands. With one exception (Merkel's disks), the transducer region of all these receptors is found on the nerve terminal. In most cases, particularly when small afferent fibers are involved, the nerve terminal lies within the skin, is free of any obvious specialized structure, and is thus called a "free nerve ending." All pain and temperature receptors and a majority of touch receptors are of this type. However, the most important touch receptors are surrounded by connective tissue encapsulations that are used to transmit the mechanical deformation of the skin to the nerve terminals where the transducer membrane is located.

Classification of Sensory Neurons

Two different nomenclatures have been used for identifying the axons conveying sensory information into the spinal cord. One of these is based on conduction velocities and divides axons into A and C fibers. The A fibers are myelinated and are further subdivided into alpha (α), which are the fastest, beta (β), gamma (γ), and delta (δ), which are the slowest fibers. The C fibers are unmyelinated. The second method of dividing fibers is based on their size. The largest fibers are called group I. These have diameters between 12 and 20 microns. Those with diameters between 6 and 12 microns are called group II, while those with diameters between 1 and 6 microns are called group III. These are all myelinated fibers. The unmyelinated fibers, with axons less than 1 micron in diameter, are called group IV axons. Because of the relationship between fiber diameter and conduction velocity discussed in Chapter 3, there is no difficulty converting from one naming scheme to the other. A sensory nerve usually contains about 5 times as many unmyelinated as myelinated axons, although the number of unmyelinated fibers is much less in nerves innervating the face and hands.

There is no particular virtue in either classification scheme that would make it more useful than the other. However, certain conventions have developed in the naming of particular nerves. For example, the sensory fibers associated with muscles are named as group I or II axons depending on their size, while those from the skin are called A or C depending on whether or not they are myelinated. The classification used to describe the axons associated with each of the receptors is indicated as they are discussed.

Although histologists have identified a large number of sensory receptors based on the morphology of the capsules surrounding the sensory axon, only a few have been found to have physiological significance. These are pacinian corpuscles, Ruffini nerve endings, and Meissner corpuscles (Fig. 7-1).

Pacinian Corpuscles

The capsules that form pacinian corpuscles are onion-like structures about 1.0 mm long and 0.7 mm wide. They are constructed of a series of concentric sheets (lamallae) separated by fluid. Large (group II or Aβ) myelinated sensory axons enter the capsules. Once inside, the axons lose their myelin covering and travel to the other end of the capsule as unmyelinated fibers. In addition to their presence in the skin, pacinian corpuscles are found in association with the long bones of the limbs and within the mesentery. Pacinian corpuscles are involved in the perception of vibration.

Ruffini Nerve Endings

The capsules of Ruffini endings are about 1 to 2 mm long by 150 microns wide. The capsules are filled with fluid and contain a number of connective tissue fibrils that are continuous with the connective tissue of the skin. As the sensory axon (group II or Aβ) enters a capsule, it loses its myelin sheath and branches extensively. The unmyelinated branches form connections with the internal connective tissue fibers, and thus deformations that occur on the skin surface can be transmitted to the nerve terminals. Ruffini endings generate the sensory information required for our perception of pressure.

The Meissner Corpuscle

The final type of encapsulated ending described here is the Meissner corpuscle. These receptors

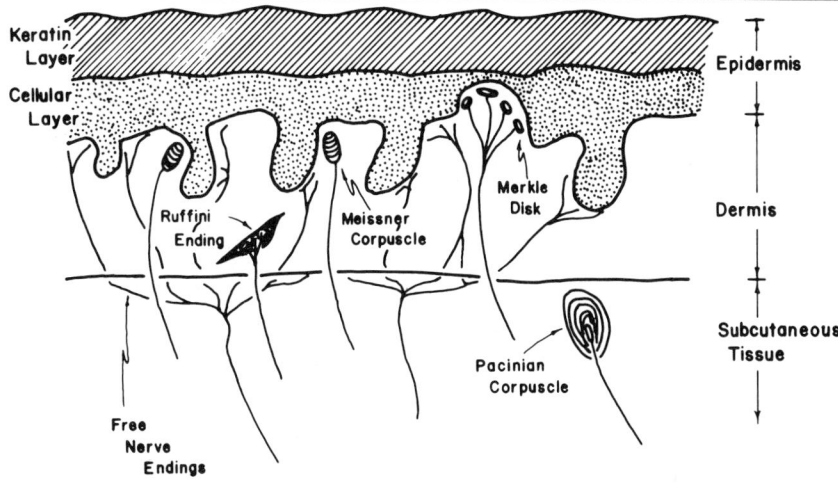

Fig. 7-1. Types of receptors innervating the skin.

are much smaller than pacinian corpuscles or Ruffini endings; each is only 30 microns wide by 150 microns long. The connective tissue lamellae that form Meissner corpuscles are densely packed and separated by an unidentified ground material. The myelinated nerves entering Meissner corpuscles branch profusely and make close contact with the laminae forming the capsule. The sensory information derived from Meissner corpuscles is used to encode the velocity of a cutaneous stimulus.

The Merkel's Disk

The most unusual of the cutaneous receptors are Merkel's disks. They are found on the border of the epidermis and dermis, making them the most superficial of the specialized receptors. The disks are formed by a group of specialized tactile cells that are in close contact with the surrounding epithelial cells. The skin area above the Merkel's disk is elevated into "touch spots" about 0.5 mm in diameter. A myelinated nerve (group II) enters the disk, branches, and makes contact with the tactile cells. Since the tactile cells contain synaptic vesicles, it is presumed that they communicate with the primary afferent neuron by synaptic transmission. Most likely, Merkel's disks are used to convey information about the location of a touch stimulus.

All of these cutaneous mechanical receptors respond to mechanical distortion of the skin. From a functional point of view, they can be separated on the basis of their adaptation characteristics and the size of their receptive fields.

FUNCTIONAL CHARACTERISTICS OF TOUCH RECEPTORS

Encoding of Vibration

Both pacinian and Meissner corpuscles are rapidly adapting receptors. However, their functions are not the same. Pacinian corpuscles detect vibrations. When a vibratory stimulus with a frequency between 50 and 900 oscillations per second is applied to the skin, the pacinian corpuscle generates one action potential per cycle. This is the same range of vibratory frequencies that can be sensed by human observers, indicating that the sensory code for vibration is based on the frequency of firing in the afferent neuron.

This coding mechanism is an exception to the general rule described in the Chapter 6. Recall from that discussion that the frequency of firing in a sensory neuron encodes the intensity of a stimulus or its rate of application depending on whether the fiber is of the tonic or phasic type. In the pacinian corpuscle, the firing frequency participates in encoding the quality of the stimulus, namely its frequency of vibration. Chapter 10, Hearing, describes how the quality of a tone can also be encoded by a labeled line and frequency code.

Encoding of the Rate of Stimulus Application

The Meissner corpuscle is utilized to encode the rate of stimulous application. As the speed of stimulus application is increased, the frequency of discharge in the neuron innervating the Meissner corpuscle increases. When the stimulus reaches its maximum strength and is no longer increasing, the Meissner corpuscle stops firing.

Encoding of Intensity and Location

The Ruffini ending and Merkel's disk are slowly adapting receptors. When the skin is stretched, the Ruffini ending fires at a constant rate that is proportional to the amount of distortion. The Merkel's disk also fires at a frequency related to the magnitude of the mechanical stimulus, but unlike the situation occuring with the Ruffini ending, there is a burst of action potentials indicating the initiation of the stimulus.

Although both of these receptors are slowly adapting, they transmit different aspects of a touch stimulus to the brain. The intensity of the stimulus is most likely encoded by the Ruffini ending, because its firing rate varies very little over time. In addition, the perception of pressure is probably elicited as a consequence of the sensory signal transmitted to the brain by the Ruffini ending. Merkel's disks, on the other hand, are more involved in signalling the location of a stimulus because of their small receptive fields.

Receptors with small receptive fields are capable of providing precise information about the location of a stimulus applied to the skin. Pacinian corpuscles and Ruffini endings have receptive fields extending over a wide area of the skin. By contrast, the receptive fields of Merkel's disks

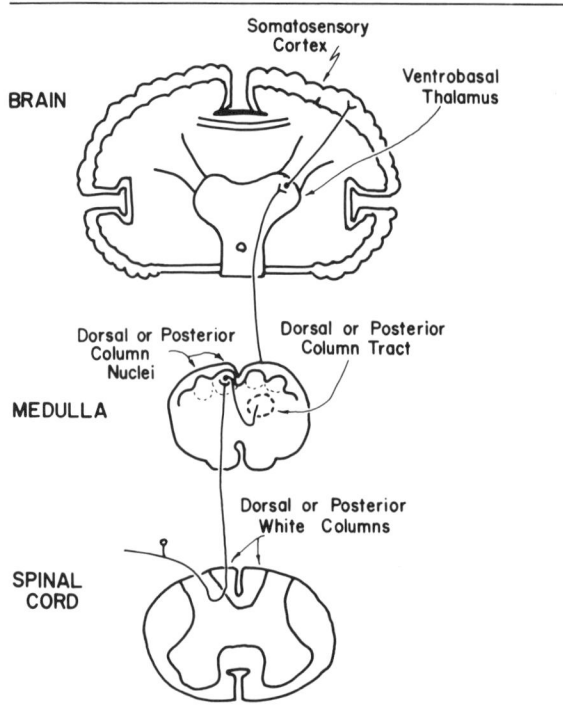

Fig. 7-2. The large afferent neurons from the skin enter the spinal cord through the dorsal root and ascend in the posterior white columns to the posterior column nuclei in the medulla. Secondary fibers from these nuclei project to the thalamus as the medial lemniscus. From the thalamus, tertiary fibers travel through the internal capsule to the somatosensory cortex of the parietal lobe.

and Meissner corpuscles are limited to a small area. This is especially true on the fingertips, where the receptor density is about 200 per cm^2.

The Sensory Pathway

The neural pathway transmitting the information encoded by these receptors is fairly simple (Fig. 7-2). The large afferents from the encapsulated endings enter the spinal cord through the dorsal root, ascend in the posterior or dorsal columns, and terminate in the dorsal column nuclei within the caudal medulla. From these nuclei, secondary neurons cross to the other side of the brain stem and travel within the medial lemniscus to the ven-

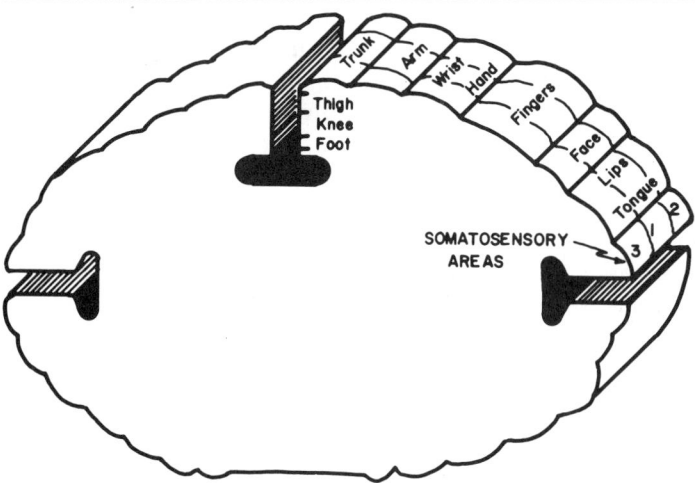

Fig. 7-3. The somatosensory cortex is topographically organized. Those areas of the skin from which the most precise sensory information can be obtained, such as the fingertips, have the largest amount of cortical area devoted to them.

tral posterolateral nuclei of the thalamus. Tertiary neurons from the thalamus then project through the posterior limb of the internal capsule to the somatosensory cortex of the parietal lobe, located just behind the central sulcus.

A single peripheral nerve contains afferents from a fairly circumscribed area of the skin. However, the sensory neurons enter the cord through several adjacent dorsal roots. As a consequence, there is a significant amount of overlap between the area of skin innervated by neighboring dorsal roots. Nonetheless, each dorsal root sends sensory fibers to a specific region of the skin called the *dermatome*. Damage to a dorsal root causes loss of sensation from that dermatome. Thus, during a neurological examination the extent of injury to the spinal cord can be ascertained by mapping out the boundaries of the sensory deficit.

Topographic Organization

The precise spatial organization characterized by the dermatomes is maintained throughout the sensory pathway. That is, the neurons conveying information from different areas of the skin are segregated from each other. In the spinal cord and medulla, the neurons originating in the face and arms are located laterally and below those originating from the legs and trunk. In the brain stem, the pathway crosses to the contralateral side, and the face and arm areas become medial to those of the lower body. This orientation shifts once again in the thalamus so that, as illustrated in Figure 7-3, the cortical projections from the face and arms are lateral to those from the trunk and legs.

This type of neuronal organization, in which specific sensory information is projected to separate anatomical regions, is called *topographic representation*. In this case, since cutaneous sensation is being conveyed, it is referred to as *somatotopic representation*. The map of the skin areas represented on the cortical surface (Fig. 7-3) is called a *sensory homunculus*.

Although not indicated in Figure 7-3, the cortical representation indicates both the location of the receptor field and the type of information conveyed by the primary afferent fiber. For example, cortical neurons receiving information from the slowly adapting receptors are located in the frontal region of the sensory cortex, while those receiving projections from the rapidly adapting receptors are located posteriorly. Another important aspect of cortical representation is that those areas of the skin with a high density of innervation have the greatest number of cortical neurons reserved for them. Because of this, areas such as the lips, tongue, and fingers, which convey a great deal of sensory information, occupy a much larger area of the sensory cortex than do the trunk and legs, which convey relatively little sensory information.

Two-Point Discrimination

The density of receptors and the size of their receptive fields determines how well a stimulus applied to the skin can be localized. The most precise way of testing this capability is to determine the "two-point threshold." This is the smallest distance that two tactile stimuli can be placed from each other and still be distinguished as separate stimuli. As can be inferred from the high density of receptors with small receptive fields (and the correspondingly large cortical area devoted to them), the two-point threshold on the fingertips is much smaller (2 mm) than the two-point threshold on the forearm (20 mm) or back (40 mm). The two-point threshold is made even smaller because of the lateral inhibition that occurs within the pathway carrying tactile information to the cortex.

As diagrammed in Figure 7-4, when a stimulus is applied to a receptive field of one afferent fiber, branches of that neuron act to inhibit the activity generated in neighboring neurons. The general effect of this inhibition is to "sharpen" the ability to locate the point of stimulus application. When two stimuli are applied, the ability to distinguish them as separate is enhanced, because the activity generated in receptors located between the two stimuli is reduced or eliminated. The ability of lateral inhibition to enhance spatial acuity is discussed again when the visual system is described.

Fig. 7-4. *Two-point discrimination is aided by lateral inhibition. The two probes cause equal amounts of excitation in all three of the neurons illustrated. However, the information conveyed by the middle fiber is inhibited. As a result, the information received by the somatosensory cortex evokes a sensation of two distinct stimuli.*

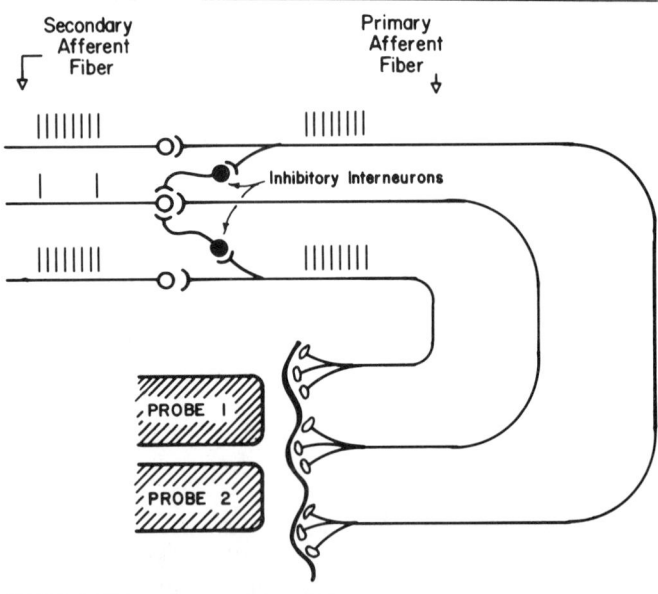

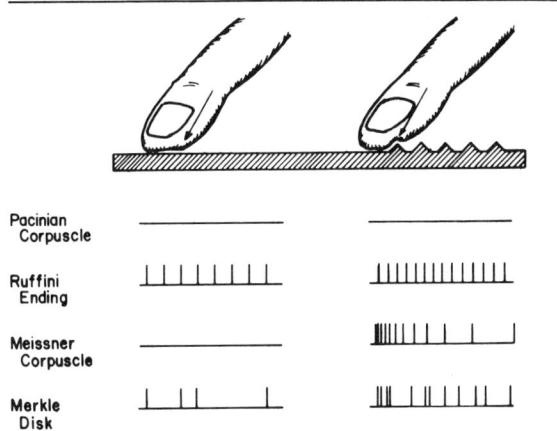

Fig. 7-5. Sensory perceptions are derived from a variety of sensory receptors. The perception of a smooth surface results from the activity of tonic receptors. Quite a different perception results when the finger moves over a rough surface and the ridges activate the phasic receptors as well.

Synthesis of Sensory Information

Although each of the receptors has been described separately here, they all act together to encode the information necessary for cutaneous mechanical perceptions. Figure 7-5 illustrates the activity generated in each of the receptors when a finger is passed over a complex spatial pattern. While the finger is over the flat portion of the board, both of the slowly adapting receptors (the Ruffini ending and the Merkel's disk) are active. The Merkel's disk locates the part of the finger in contact with the board and the Ruffini ending determines how much pressure is being applied. As the first part of the pattern is reached, there is a burst of activity from the rapidly adapting Meissner corpuscle, indicating the steepness of the ridge. The increased activity in the Ruffini ending due to the added pressure on the finger encodes the height of the pattern. At all times the cortex is aware of the stimulus location because it "knows" which one of the receptors with small receptive fields (Meissner corpuscle or Merkel's disk) is firing.

Thus, in considering the encoding of sensory information by the skin, it is necessary to know the specific function each of the cutaneous receptors serves. But it is equally important to appreciate that a mosaic of sensory information is provided by the activity of all the receptors acting in parallel. It is the synthesis of all this information by the cortex that enables us to obtain the complex sensory experiences that characterize our tactile perceptions.

PAIN RECEPTORS

Pain is unique among the cutaneous sensations because, in addition to identifying a particular type of sensory stimulus, it produces a powerful emotional response and has a strong effect on behavior. Its most important function is to protect an organism from continuing behaviors that are harmful. For example, the pain caused by a blister causes an individual to stop running before more serious tissue damage results, and similarly, the pain associated with a disease causes us to cease our activity and seek help from a physician. In this regard pain is a helpful if unpleasant sensation.

However, if the pain sensation continues after its protective function has been served, the pain can become extremely debilitating. In cases of chronic pain, the emotional effects of pain become the dominant force in an individual's life, preventing normal activities from being carried out and eliminating all sources of pleasure. Individuals suffering from chronic pain are therefore in desperate need of relief. Unfortunately, the therapeutic measures currently available are often inadequate. Although pain medications (such as morphine) are helpful for a short time, they result in a large number of deleterious side effects and can lead to drug addiction. Surgical procedures, in which the components of the anatomical pathways carrying pain are lesioned, usually do not provide permanent relief and often leave the patient with neurological deficits. Even worse, chronic pain syndromes can exist without any obvious physical injury to account for them.

Under these conditions, the patient may be accused of malingering or being hypochondriacal, and thus in addition to suffering from excruciating pain may lose the support of his or her physician and family. Much more information about the causes of chronic pain must be gathered before its cause can be understood and its presence eliminated.

The following paragraphs describe the neurophysiological mechanisms underlying the sensation of pain and indicate some of the ideas that have been proposed to explain the development of chronic pain and its treatment.

Activation of Pain Receptors

Pain is a complex perception that includes both the sensation evoked by strong mechanical, thermal, or chemical stimuli and the emotional and psychological results of the pain sensation. For this reason, it is common to identify the receptors for pain as *nociceptors*, to indicate that they are activated by noxious stimuli and to separate the sensory aspects of the perception of pain from the emotional attributes of the pain experience. However, the discussion here uses the terms *pain receptor* and *nociceptor* interchangeably.

The receptors for pain are found on the ends of small myelinated (A δ) and unmyelinated (C) sensory fibers. There are no specialized encapsulations covering them, so they are referred to as free nerve endings. Some of these receptors (particularly the myelinated A delta fibers) respond only to strong mechanical stimuli, while the others (the C fibers) respond to mechanical, thermal, and chemical stimuli. Because they respond to different types of stimuli, the small afferents are called *polymodal nociceptors*. In all cases, the fibers that respond to pain stimuli are specialized for that purpose. The other free nerve ending receptors in the skin (and elsewhere within the body) that produce sensations of touch or temperature do not cause pain sensations, no matter how intensely they are stimulated.

The mechanism by which strong stimuli activate the pain receptors is not known. However, there is evidence to support the idea that the pain stimulus causes tissue damage and that some substance released from the interior of the damaged cells stimulates the receptors. This substance may be potassium, histamine, or a polypeptide such as substance P or bradykinin. Prostaglandins may also contribute to the sensation of pain, since analgesic agents such as aspirin act by inhibiting the synthesis of prostaglandins.

The Dual Sensations of Pain

A typical pain sensation consists of two separate perceptions. For example, if a pin pierces the skin, a sharp, pricking type of sensation is perceived. This is a well localized and well defined sensation that gives rise to protective reflexes (such as withdrawal of a hand from a hot stove) and an increase in sympathetic nervous system activity (causing, for example, an increase in cardiovascular activity and mobilization of blood glucose stores). Soon afterwards (0.5 to 1.0 second), a second sensation begins. This is not as well-localized and is described as aching, throbbing, and unpleasant. If severe, the painful sensations are accompanied by nausea, sweating, and a reduction in blood pressure. The initial sharp pain is evoked by activity in the myelinated fibers, while the second dull pain is caused by the firing of the unmyelinated fibers.

The Pain Pathways

The temporal separation of the two pain sensations can be explained on the basis of the conduction velocity and neuronal pathway by which the two types of pain fibers reach the cerebral cortex. The propagation velocity of the small myelinated fibers (10 m/s) is much greater than that of the unmyelinated C fibers (less than 1 m/s), so the information conveyed by the former fibers reaches the spinal cord first. Both types of pain fibers terminate within the upper part of the dorsal horn, in the marginal zone and substantia gelatinosa (Fig. 7-6). However, their pathways to the brain are different.

The fibers responsible for producing the initial pain sensation enter the spinal cord in the dorsal

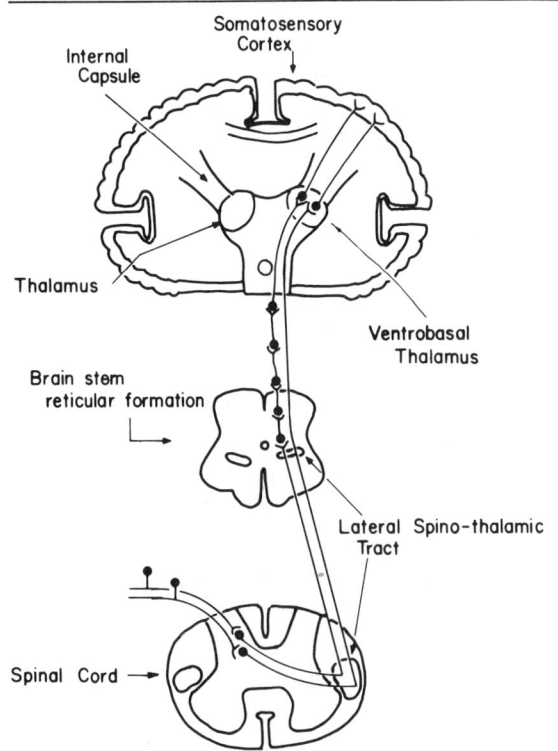

Fig. 7-6. *Most of the myelinated pain fibers entering the spinal cord in the dorsal root synapse within the superficial layers of the dorsal horn. Secondary fibers then cross to the contralateral cord through the anterior spinal commissure and ascend as the lateral spinothalamic tract. The unmyelinated pain fibers follow the same course to the contralateral side of the spinal cord but ascend in a less well-defined, diffuse pathway to the brain stem reticular formation.*

roots and synapse immediately with interneurons in the upper portion of the dorsal horn. These interneurons then synapse ipsilaterally with tertiary neurons whose axons cross to the contralateral cord and ascend to the thalamus as the anterolateral spinothalamic tract. Projections from the thalamus then convey the information about the pain stimulus to the cortex. Pain sensations are transmitted fairly rapidly and precisely through this system because it contains relatively large fibers, has few synapses, and maintains a precise topographical representation from spinal cord to cortex.

The pathway carrying the delayed pain to the brain is not as well characterized. It is generally described as a spinoreticulothalamic pathway and is referred to as a paleospinothalamic tract to indicate that it was present, from an evolutionary point of view, before the spinothalamic tract developed. The spinoreticulothalamic pathway is slow, requires many synapses to reach the cortex, and does not have a precise topographical representation. It is bilateral and appears to include a number of different ascending tracts.

Functionally, the information carried within the paleospinothalamic tract reaches the brain after information conveyed through the neospinothalamic tract, and thus can account for the presence of a delayed, poorly defined pain sensation. In addition, the projection of the paleospinothalamic tract to the reticular formation and from there to areas of the brain associated with emotion is offered as an explanation for the unpleasantness associated with the delayed pain.

Another aspect of the "older" spinoreticulothalamic system that is important from a clinical point of view is that this pathway is difficult to eliminate by surgical sections of the spinal cord, such as the anterior quadrant cordotomies carried out for the relief of chronic pain. This difficulty arises because the pain afferents enter the spinal cord over many different segments and form multiple pathways to the brain, not all of which are in the anterior quadrant.

In addition, the removal of the afferent fibers in a pain pathway may cause the more central neurons to become hypersensitive (in a manner similar to that occurring in the denervation supersensitivity described earlier in Chapter 4) and thus to maintain the pain perception by firing spontaneously. It has also been found that a large number of fibers responsible for eliciting the diffuse type of pain enter the spinal cord through the ventral roots, making dorsal root sections ineffective in eliminating pain. The nature of these extensive anatomical pathways is such that, although surgical intervention is occasionally successful in eliminating pain for a short time, it almost never produces lasting relief from pain.

Referred Pain

Often, disease or injury of internal organs produces a sensation of pain that appears to arise from some area of the skin. This is called *referred pain*. It is different from the projected pain that occurs when neurons within the pain pathway are stimulated. In the latter situation, the pain sensation is felt at the peripheral site from which the pain fiber originated. It occurs at that site because, as indicated earlier in the discussion of the labeled line mechanism of sensory coding, stimulation of the neuronal pathway at any point evokes the sensation encoded by that neuron. For example, compression of the sciatic nerve bundle can cause a painful sensation that appears to come from the legs.

The explanation for referred pain is not as well understood as the anatomical organization of the pain pathway, but seems to be related to it. Pain sensations thought to be caused by injuries or diseases of the viscera cannot be precisely localized, because there is little or no representation of the viscera within the sensory cortex. Instead, when pain afferents from the viscera are stimulated, they cause excitation of the same interneurons that are used by cutaneous pain fibers. When these interneurons are stimulated, they cause the localization of the stimulus to be referred to the cutaneous site. Referred pain is an important diagnostic aid because it is an early and often the only sign of ongoing visceral disease. The best known example of this is the common observation that cardiac pain appears to come from the chest and arm, and not from the heart (Fig. 7-7).

Pain Originating Within the Nervous System

One of the major clinical problems associated with pain is that pain sensations can exist long after their initial causes have disappeared. For example, after an amputation there is often a subjective feeling that the limb is still there. This is called the *phantom limb phenomenon*. Occasionally, the sensations appearing to originate in

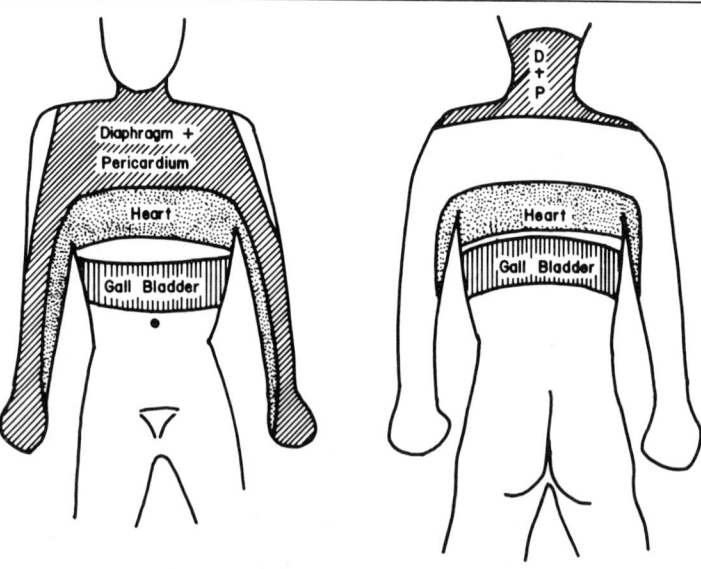

Fig. 7-7. Surface areas to which visceral pain is referred.

the amputated limb include that of severe pain. How this phantom limb pain develops is not known. However, it continues to exist even after all the injuries associated with the amputation have healed. To explain this and other cases of chronic pain occurring without any obvious cause, a number of theories have been proposed.

The most influential of these theories is the gate-control theory, which proposes that there are specific pain neurons within the spinal cord. Their firing rate is controlled by a gate that, in turn, is controlled by two opposing synaptic inputs. Large afferent fibers from cutaneous receptors close the gate, while small afferent pain fibers open it, causing pain fibers to fire. The gate-control theory explains why rubbing an injury causes the pain to be reduced, since rubbing excites many large sensory neurons that then act to close the gate and relieve the pain.

Some of the specifics of the gate-control theory have been found to be incorrect. However, the concept underlying its mode of action, which is that pain fiber input to the brain can be reduced or eliminated by the activity of other neuronal pathways, has been substantiated and used to develop therapies for the treatment of chronic pain. For example, surgeons have implanted electrical stimulators on the dorsal columns of patients suffering from intractable pain. When the patient turns the stimulator on, the large sensory neurons within the dorsal column are activated and pain is found to be reduced. Even electrical stimulation of peripheral nerves through the skin (transcutaneous electrical stimulation or TENS) has been found to be effective in many cases. And, of course, the success of acupuncture has been attributed to the stimulation of large afferent fibers by the acupuncturist's needle.

Conversely, the gate-control theory can be used to explain the presence of pain in the absence of obvious peripheral injury. The gate controlling the production of pain is not only under the influence of peripheral ascending fibers but descending pathways as well. If these descending inhibitory pathways do not function properly, pain results from physiologic rather than anatomic causes.

Descending Inhibition of Pain

One of the pathways found to be capable of controlling the transmission of pain passes through the periaqueductal gray matter of the brain stem. When this pathway is stimulated electrically, neurons within the spinal cord that are responsible for transmitting pain are inhibited. The axons of this descending inhibitory pathway use serotonin (5-hydroxytryptamine) as their neurotransmitter and interestingly, a group of neurons within this pathway is activated by morphine and other opiates. Thus, it appears that one of the major ways in which morphine produces its analgesic effects is by activating a descending pain-inhibitory pathway.

These observations, coupled with the discovery of opiate binding sites within the central nervous system and the isolation of endogenous substances (the endorphins and enkephalins) within the brain that are capable of binding to them, have opened up an entirely new avenue to explore in the treatment of chronic pain. Many endogenous morphine-like substances have now been isolated from the central nervous system and chemically characterized. They and a large number of synthetic analogs possess analgesic properties similar to morphine. Unfortunately, although some of them are much more potent than morphine, they all tend to lose their effectiveness with repeated doses and produce the same sort of drug dependency that morphine does.

Nonetheless, there is great hope that an analog to the naturally occurring morphine-like compounds will be found that possesses a potent analgesic effect without causing any of the severe side effects that now characterize these drugs.

THERMO RECEPTORS

Thermoreceptors, responding to increases (warm receptors) and decreases (cold receptors) in temperature, are located on the free nerve endings of small myelinated (A δ) and unmyelinated (C) sensory fibers. The receptive field for these fibers is rather small, and cold and warm fiber receptive fields do not overlap. An area of skin usually receives about 5 to 10 times more cold fibers than warm fibers.

Cold and Warm Receptors
Both cold and warm fibers fire at a steady rate to a sustained skin temperature. But, as illustrated in Fig. 7-8, the temperature range over which they fire is different. Cold fibers fire at a maximum rate when the skin temperature is maintained at about 27°C, whereas warm fibers reach a maximum rate of constant firing when the skin temperature rises to 40°C. Cold fibers continue to fire until the skin temperature falls to below 10°C. They also stop firing when the skin temperature reaches 40°C, but interestingly, start firing again when the skin temperature rises above 43°C. Warm fibers fire over a much narrower range of temperatures. They begin to fire at approximately 27°C and then abruptly cease firing when the skin temperature becomes approximately 42°C.

The tonic rate of firing provides the central nervous system thermoregulatory centers with information about skin temperature. However, sensory perception of skin temperature is much more closely related to the phasic changes in firing rate that occur when the skin temperature changes.

Cold fibers always increase their firing rate when the temperature is decreased and decrease their firing rate when the temperature is increased. The firing frequency achieved during a temperature change depends on the initial skin temperature and the amount by which the skin is cooled. Warm fibers behave in an opposite fashion, increasing their firing rate in response to an increase in temperature and decreasing their firing rate when the temperature falls. In both the cold and warm fibers the increase in firing rate is a phasic response. After a few seconds, the fibers adapt to a firing rate that depends on the actual temperature of the skin, as indicated in Figure 7-8.

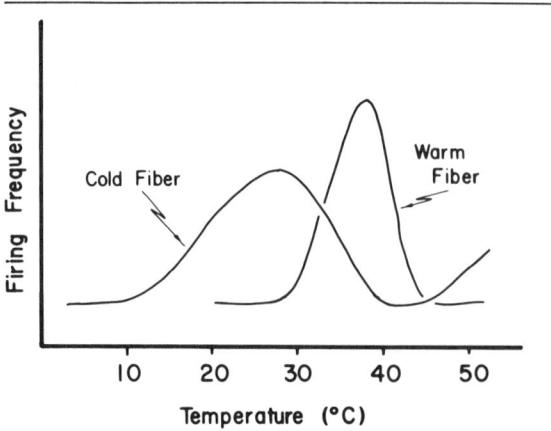

Fig. 7-8. Temperature range over which the cold and warm fibers are active. The cold fibers stop firing at approximately 40°C but begin firing again when the temperature reaches 45°C, producing the perception of paradoxical cold.

Thermal Sensations
The perceptual qualities of temperature sensation can be explained by the way in which the thermoreceptors respond to a change in temperature. For example, if one enters a warm bath or a cool lake there is an immediate sensation of warmth or cold, but after a short time the feeling of temperature disappears. The elimination of a temperature sensation after spending a few minutes in a constant temperature environment probably reflects the adaptation of the peripheral thermoreceptors.

The situation is complicated by the thermoregulatory mechanisms within the hypothalamus. These brain centers are responsible for maintaining the core temperature of the body at about 37°C. They respond to an increase in temperature by dilating peripheral blood vessels and increasing the rate of sweating. Decreases in temperature are met by constricting of peripheral blood vessels and shivering. These physiological responses are accompanied by feelings of warmth, cold, comfort, and discomfort, feelings that can interfere with the sensory perceptions produced by the thermoreceptors.

Thus, when exposed to a constant temperature below 30 °C there is a persistent feeling of cold that is probably related more to the thermoregulatory response of the brain than to the neural discharge of cold fibers. Similarly, temperatures above 35 °C cause a constant feeling of warmth. At temperatures between these two values, there is no sustained temperature perception. Correspondingly, the thermoregulatory adjustments required to maintain a constant body temperature are at a minimum within this temperature range.

Temperatures below 15 °C and above 45 °C produce a feeling of pain. The pain sensation is caused by specific pain fibers, not by the discharge of thermoreceptors. In fact, the warm fibers do not fire at the high skin temperatures producing pain, but as indicated earlier, the cold fibers begin to fire again at these temperatures. This produces the sensation of paradoxical cold in which there is a distinct feeling of coolness followed by a burning, painful sensation.

The ability to perceive a change in temperature varies with the initial adapting temperature of the skin. Thus, when the skin is held at a temperature below 30 °C, a sudden decrease in temperature by less than 0.2 °C can be detected. At a temperature above 35 °C the skin must be cooled by almost 1 °C before it can be detected. A similar phenomenon occurs for the detection of temperature increases. These are much easier to detect when the skin is adapted to a warm temperature than when it is adapted to a cold temperature. This variability in the detection of temperature changes is mirrored by the responsiveness of the thermoreceptors to sudden changes in skin temperature.

Because cold fibers increase their firing rate whenever the skin is cooled, the sensation of cold can be evoked even though the skin is never exposed to a "cold" temperature. For example, if a hand is placed in a pail of water heated to 35 °C and then moved to a pail containing water at 32 °C, the second pail of water feels cool. In contrast, if the hand is first placed in a pail of water at 30 °C and subsequently moved to the pail containing the 32 °C water, the pail of water is perceived as warm. This presumably occurs because the warm receptors increase their rate of firing in response to the increase in temperature. Thus it is the direction of temperature change rather than its absolute value that determines whether a stimulus is perceived as cold or warm.

8 Taste and Smell

More than 2,000,000 American adults suffer from some sort of gustatory (taste) or olfactory (smell) sensory disorder, and their illnesses can have severe consequences. For example, with anosmia (absence of the ability to smell) an individual is not able to detect a gas leak or fire, while ageusia (absence of the ability to taste) leaves an individual unable to recognize spoiled or adulterated food. Even more debilitating are those diseases that produce unpleasant alterations in the way taste and smell stimuli are perceived (dysgeusia or dysosmia, respectively). These conditions not only diminish the enjoyment of eating and drinking, but can greatly affect the nutritional status of the patient. It is very important for a physician to recognize these disorders because they can be a sign of brain tumors, temporal lobe epilepsy, or disturbances of the endocrine, kidney, or gastrointestinal systems. Thus, an evaluation of these sensory systems is an important part of a thorough neurological examination.

The role of taste and smell in normal behavior is not well understood, but may be of great significance. The foods we choose to eat are related to how pleasant they taste or smell, and so the sensory information obtained from our gustatory and olfactory receptors influences our diet and affects our nutrition. It is well known that animal behavior can be greatly modified by airborne odors called *pheromones*. Whether these substances affect human behavior is controversial, but there is increasing evidence that specific odorous compounds are released from humans in relation to sexual activity, stress, and disease. Certainly the deodorant and perfume industries have shown a great interest in developing products to mask naturally released odors and replace them with odors designed to elicit the appropriate behavior in others.

The following section describes the receptors

involved in the detection of taste and smell stimuli and the process of encoding this information for transmission to the central nervous system. The chapter concludes with a description of some of the techniques used to assess the functional status of the chemosensory apparatus.

TASTE

The Taste Pathways

The stimulus for taste is a chemical dissolved in the fluid bathing the sensory receptors. The sensation of taste is conveyed from receptors on the tongue to the brain by the VII (facial) and IX (glossopharyngeal) cranial nerves. The chorda tympani nerve (a branch of the facial nerve) innervates the front two thirds of the tongue, while the glossopharyngeal nerve innervates the back of the tongue. Both nerves project to the medulla from which second-order neurons arise. These travel to the thalamus, which in turn relays the sensory information to the cortex, where conscious sensation presumably arises.

There is another taste pathway in which activity of the medulla is transmitted to the pons and then to the hypothalamus and amygdala. This pathway is thought to convey information used to generate activity related to eating. For example, nutritious foods such as carbohydrates evoke feeding behavior, while naturally occurring poisons such as alkaloids (quinine or strychnine) produce reflexes that cause the food to be rejected. These responses occur independently of the taste perceptions associated with these compounds. In addition, certain compounds can cause different effects on appetite, depending on the nutritional status of the individual. For example, in cases of salt deficiency (which can occur in a disease of the adrenal gland called Addison's disease) patients eat food with concentrations of salt that would be totally unpalatable to normal individuals.

Because so little is known about the role of the central nervous system in taste perception, we

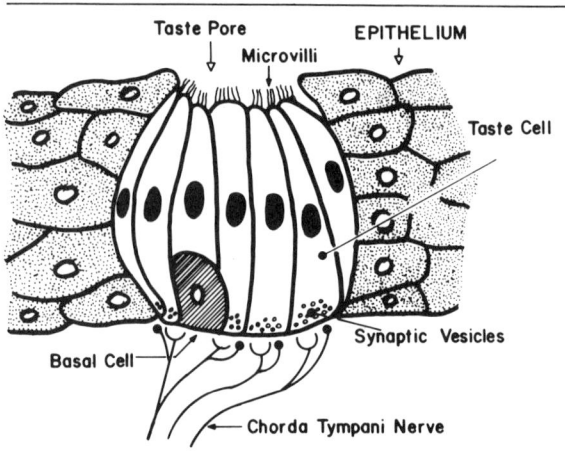

Fig. 8-1. *The taste bud is composed of 40 to 50 taste cells. Each taste cell contains a number of microvilli on which the taste receptors are found. New taste cells are continuously being formed from the basal cells.*

shall confine our discussion to the peripheral mechanisms involved in sensory encoding.

The Taste Receptors

The receptors for taste are found on specialized cells called *taste cells* (Fig. 8-1). The taste cells are grouped together in a globular structure called the *taste bud*. The taste buds are contained in protrusions of the tongue called *fungiform* and *circumvallate papillae* (Fig. 8-2). Several hundred fungiform papillae, each containing 3 to 5 taste buds, are found scattered over the anterior two thirds of the tongue. At the back of the tongue is a row of about 10 circumvallate papillae. These each contain a few hundred taste buds.

Taste buds are about 100 microns in diameter and contain 40 to 50 taste cells (see Fig. 8-1). These cells are modified epithelial cells that are being continuously differentiated from cells found at the base of the taste bud. Once formed, the cell migrates to the surface of the taste bud, where it is sloughed off. The life span of the cell is approximately 8 to 10 days.

The taste cells are arranged in a circle around

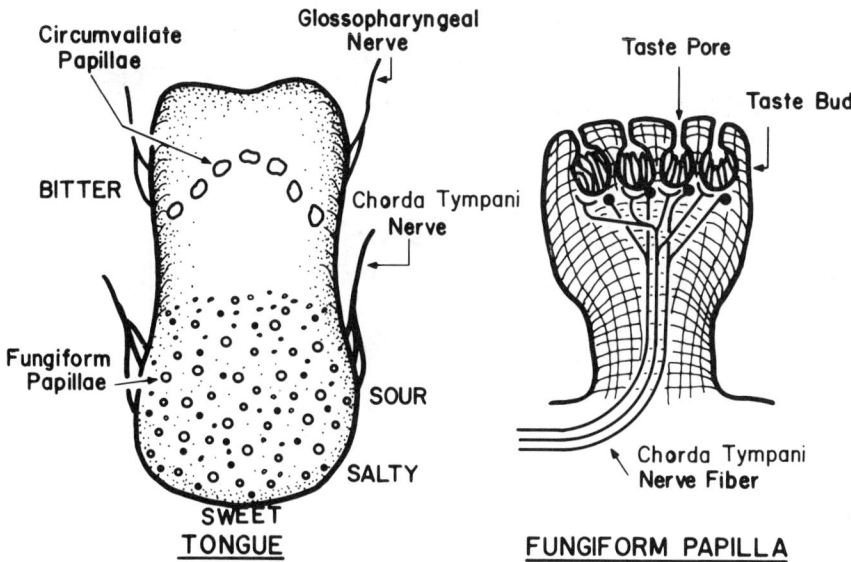

Fig. 8-2. The taste buds are located on the dorsal surface of the tongue. Some regions of the tongue are more sensitive to one of the primary taste sensations than to the others. Each of the several hundred fungiform papillae contains 3 to 5 taste buds.

a taste pore through which saliva, in which the taste stimulus is dissolved, can enter the taste bud. Microvilli, found on the apical ends of the taste cells, project into the pore. It is assumed that the microvilli contain the receptor sites for gustatory stimuli. The basal end of the cell is filled with synaptic vesicles. This situation is similar to that occurring for receptor cells in Merkel's disks and in the ear. There too the sensory cell is a modified epithelial cell rather than a nerve ending, and the communication between the receptor and the primary afferent neuron is by synaptic transmission.

The sensory nerve fiber is essential to the continuous differentiation of new taste cells. If the sensory nerve is transected, the taste buds degenerate. Following reinnervation of the papillae, the taste buds reform within a few days. The mechanism by which the sensory nerve controls the growth of the taste bud is not known, although there is good experimental evidence to indicate that a trophic substance, transported down the nerve axon, is responsible for the development and maintenance of the taste buds. The role of efferent neurons in maintaining the integrity of their postsynaptic cells has already been discussed. In this case, it is the afferent fiber that is necessary for the survival of the presynaptic cell. These examples of the trophic function of neurons indicate that their physiological role is not limited to information processing.

The Taste Sensations

In order to be effective, a gustatory stimulus has to be dissolved in saliva. It then flows into the taste pore from which it can bind to the receptor sites on the microvilli of the taste cells. The stimulus-receptor complex next causes the taste cell to depolarize and release its neurotransmitter. The nature of the receptors is controversial.

Classically, there were thought to be four primary taste stimuli: sweet, sour, bitter, and salty. It was assumed that there was one receptor molecule corresponding to each of these primary sensations. Supporting this view is the observation

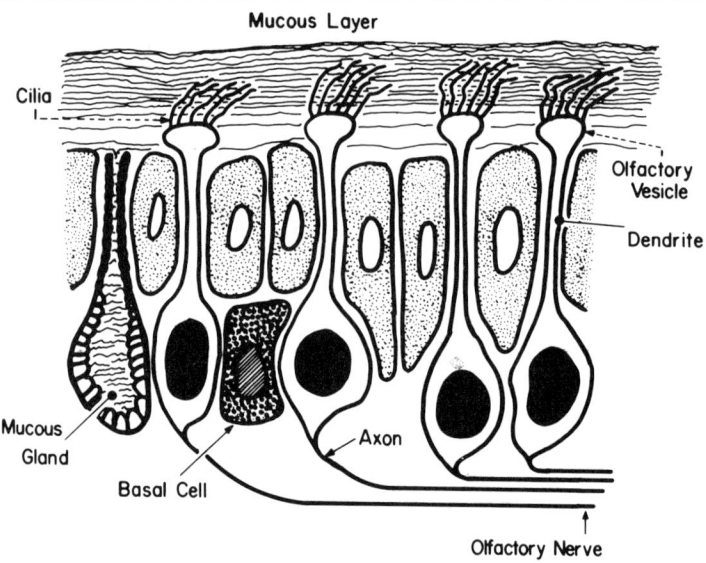

Fig. 8-3. The olfactory receptors are located on the cilia projecting from an expansion of the olfactory (first cranial) nerve terminals, called the olfactory vesicles. New olfactory nerve fibers may be formed from the basal cells within the olfactory epithelium.

that papillae in different regions of the tongue are primarily responsive to only one of the four taste sensations. As Figure 8-2 illustrates, the front of the tongue contains papillae that respond only to sweet-tasting stimuli, while the back of the tongue responds to bitter-tasting stimuli. Along the side of the tongue, the anterior region is most sensitive to salty stimuli and the posterior region is most sensitive to sour stimuli. Further support for the concept of specific receptors is the fact that each of the four sensations of taste appears to be evoked by a different class of chemical compounds. For example, sugars are sweet, acids are sour, salts (NaCl) are salty, and alkaloids (such as quinine) are bitter.

The situation became more complicated when recordings of electrical activity from single sensory fibers innervating the taste buds were obtained. These neurons were found to respond to a variety of stimuli applied to the tongue. Each sensory axon appeared to have its own unique response profile. For example, one fiber might respond to bitter, sour, and salty stimuli and another to sweet and salty stimuli. In order for the central nervous system to create a perception of the stimulus, integration of the activity in all of the peripheral fibers would be required. As discussed in Chapter 6, this type of neural encoding (spatial or temporal pattern coding) utilizes the activity in many neurons to transmit sensory information to the central nervous system. The classical view (in which each neuron encodes a specific sensation) is an example of a labeled-line mechanism of encoding. Further research is required to resolve the differences between these two theories.

SMELL

Smell, like taste, is a chemical sense in which the stimulus must be dissolved before it can react with its receptor. It differs in that the stimulus is an airborne particle that has an effect on an individual while he or she is far away from the stimulus. This feature of smell is useful in that it can direct someone toward a desired aroma or away from a smell indicating a dangerous situation. Additionally, it permits a person to reject foods on the basis of their smell before they are put in the mouth.

The Olfactory Pathway

The receptors for the sense of smell are found on the dendrites of the bipolar neurons forming the first cranial nerve (the olfactory nerve) (Fig. 8-3). The cell bodies and dendrites are embedded in a layer of tissue called the *olfactory epithelium*, which is located at the top of the nasal cavity in a 10 cm² area on either side of the nasal septum. In addition to the bipolar neurons, the epithelium contains supporting and mucous cells. The latter produce a mucous secretion that covers the epithelium. Odorous molecules that enter the nose during respiration or "sniffing" must dissolve in the mucus before they can reach the receptor sites on the olfactory neurons. The dendrites and cell bodies are isolated from each other by the supporting cells. The axons, which arise from the cell body, pass through the cribriform plate (the roof of the nasal cavity) in small bundles, and immediately enter the overlying olfactory bulb.

The olfactory bulb is a highly ordered information processing station that interprets olfactory information received from the sensory neurons and transmits it to the olfactory cortex. Synaptic contact between the olfactory nerve axons and the second order neurons occurs within a spherical region of the olfactory blulb called the *glomerulus*. The neuronal structure of the olfactory bulb is organized in layers, much like the retina. When the olfactory receptors are stimulated, the neurons within the olfactory bulb can be excited or inhibited. It is thought that the inhibitory responses are utilized to reduce extraneous odors, so that the perception of the olfactory stimulus can be enhanced. This mechanism is similar to the one discussed in chapter 7, in which lateral inhibition was utilized to enhance two-point discrimination.

Second-order neurons from the lateral olfactory tract project directly to the olfactory cortex on the ventral aspect of the frontal lobe. This is the only sensory pathway that travels to the cortex without a relay station in the thalamus. In addition to this direct pathway, neurons from the olfactory bulb reach the cortex through the thalamus.

Another important pathway travels to the hypothalamus and limbic system. This pathway is responsible for the emotional and behavioral consequences of olfactory stimulation. Interestingly, fibers from the olfactory cortex send axons back to the olfactory bulb. It is possible that these fibers are used to alter the response of the central nervous system to a particular olfactory stimulus. For example, certain food odors that are quite attractive when a person is hungry may become repulsive when he or she is full. This change in attitude toward the stimulus may be mediated by the corticifugal fibers to the olfactory bulb.

The olfactory system has an intimate connection with the hypothalamus and the limbic system, and probably plays an important role in our feeding and reproductive behavior. However, because so little is known about how the central nervous system contributes to our olfactory perceptions, the discussion here is limited to the encoding of information by the olfactory receptors.

The Olfactory Receptor

As indicated earlier, the olfactory receptor is found on the dendrites of the bipolar neurons that form the olfactory nerve (Fig. 8-3). The dendrite is about 1 micron wide and 10 to 100 microns long. At its end, it gives rise to a bulb-like structure 1 to 2 microns in diameter called the *olfactory vesicle*, from which some 20 to 30 cilia arise. The cilia, which contain the receptor sites for olfaction, pass into the mucous layer covering the olfactory epithelium.

Olfactory neurons, being exposed to the external environment, can be easily traumatized, and there is some question as to whether these cells are being continuously destroyed and if so, whether they can be replaced. Although it is not certain that regeneration of damaged cells can occur, there is evidence that certain cells within the olfactory epithelium, called *basal cells*, divide to produce new olfactory neurons. If so, it would be a unique case, since there is no evidence that regeneration of central nervous system neurons takes place anywhere else.

In order to reach the receptor sites on the olfactory cell cilia, the stimulating molecule must first dissolve in the mucous layer covering the olfactory epithelium in which the cilia are embedded. The odorous molecules must be both volatile, so that they can be carried in the air flowing through the nose, and water-soluble, so that they can dissolve in the mucus. Upon coming in contact with the cilia, the odorant binds to the receptor site (much like a taste stimulus binds to its receptor on the taste cell microvilli) and causes the cell body of the olfactory neuron to depolarize. This depolarization (receptor potential) passively spreads to the spike-generating region of the olfactory neurons and elicits a train of action potentials. As with the other sensory systems discussed previously, information pertaining to both the quality and quantity of the stimulus is encoded by the unique combination of neurons firing and the frequency of their discharge.

The Olfactory Sensation

Although the structure of the olfactory receptor has not been worked out, many attempts have been made to identify the nature of the receptor by dividing the great variety of odorous stimuli into a small number of categories. It was assumed that the receptors would have a structure complementary to the stimuli that excite them. The most commonly accepted scheme divides odors into seven categories: camphorous (moth balls), musky, floral, minty, etheral, pungent (formaldehyde), and putrid (rotten eggs).

Each of these categories is considered a primary odor, analogous to the primary colors of vision and the primary sensations of taste. In support of this view are the clinical examples of individuals who have specific anosmias ("smell-blindnesses") to one or several of these categories of stimulants. In these cases, it is assumed that one or more specific receptors must be absent from the cilia of the olfactory neurons.

However, as in the case of taste perception, there are many odors that do not fit easily into any of the categories. Also, the olfactory neurons are responsive to many different types of odors, and so it is difficult to imagine how any given neuron could be responsible for encoding only one primary odor. Thus, it is generally thought that the pattern of activity generated in all the neurons activated by a particular stimulus is utilized by the cortex to identify the odor.

The olfactory receptors are extremely sensitive to the appropriate (adequate) stimulus. For example, the compound ethylmercaptan (an organic sulfhydryl compound with a putrid smell) can be detected in concentrations as low as 0.5×10.0^{-11} molecules per cubic centimeter. Since these molecules are distributed over the entire olfactory epithelium, it has been estimated that less than 8 molecules per neuron are capable of causing a detectable response. This means that for some compounds the nose is far more sensitive than the best chemical analyzing equipment. Interestingly, the range of concentrations that the olfactory system can respond to is only about 1 to 50. That is, at concentrations about 50 times threshold, the receptors are saturated. This is a very low range of responsiveness when compared to that of vision or audition, which can respond over an intensity range of 1 to 1,000,000.

When exposed to an odorant for even a short period of time, human observers adapt to the stimulus. This loss of ability to perceive the stimulus is not due to the adaptation of the peripheral receptors, because it has been shown that neither the generator potential nor the firing of the primary afferent neuron ceases during prolonged exposure to the stimulus. This is another example of the generalization made earlier that adaptation to sensory stimulation can be a function of the brain and not of the peripheral receptors.

CLINICAL ASSESSMENT OF TASTE AND SMELL

Disorders of taste and olfactory sensation can be quite debilitating, especially when instead of simply losing taste or smell acuity the patient is left with an unpleasant or obnoxious sensation of taste or smell. In most cases of this type, an otorhinolaryngological or neurological examination does not reveal any physical abnormality that might account for the symptoms, and such patients are often referred to a psychiatrist for further evaluation. But since it is now clear that a variety of nutritional deficiencies can produce these symptoms, it is important to have an objective method of assessing the function of the chemosensory receptor systems.

The sense of smell can be evaluated using a device called an *olfactometer*, with which the examiner can pass a known quantity of air saturated with a particular odor into the nostril of the patient. The amount of air required for the stimulus to be identified is then recorded. In this way the threshold for detection of each odor primary can be found. To test for the sense of taste, a drop of either sugar, vinegar (sour), quinine (bitter), or salt is placed on the tongue and the patient asked to identify the substance. The lowest concentration of the substance that can be identified is then noted.

In both of these tests, the various substances must be presented at intervals of about one minute, so that adaptation does not interfere with the ability to make the identifications. Occasionally, a patient who is suspected of malingering (falsely claiming a disability) has to be assessed. For example, if the patient claims not to be able to smell, he or she might be asked to identify the taste of chocolate. If the sense of smell is truly absent the chocolate is identified as sweet, but if the sense of smell is present, the chocolate is properly named. Similarly, in the absence of smell, coffee is called bitter, not coffee. This test emphasizes that a great deal of what we generally call taste is really flavor, and depends as much or more on the sense of smell as it does on taste.

9 Vision

To see an object, its image must be formed clearly on the retina. What is perceived, however, is not a simple replication of the retinal image. Instead, the cortex creates a visual perception from the information transmitted to it from the retina. This information is selected so that only those aspects of the retinal image necessary for the construction of a mental image actually reach the cortex. For example, information about brightness, contrast, color, and depth is encoded by retinal neurons, while unchanging components of the image (such as an area of constant brightness or color) are ignored.

The Blind Spot

Recall that part of your retina is devoid of sensory receptors (the blind spot or optic disk) and yet as you read this book or look about the room, there are no apparent blank spots. If you now look at Figure 9-1 and follow the instructions in the legend, you can demonstrate that there is indeed a blank spot. Moreover, you can prove that the cortex fills in those parts of an image that are ignored by the retina. This conclusion is supported by the numerous reports from the clinical literature citing examples of patients with fairly large scotomas (blind areas in the visual field) who are not aware of their visual defect because the cortex forms a complete image.

This chapter discusses the processing of information by the visual system. How the eye forms an image of a stimulus on the retina, and then the transduction of a light stimulus into an electrical signal by the photoreceptors and the neural encoding of visual information by the retina, geniculate nucleus, and cortex, are discussed. Next, some of the perceptual capabilities of the visual system are described, and a neurophysiological basis for understanding them is provided. The chapter concludes with a brief description of how the functioning of the visual system can be assessed in a clinical setting.

THE EYE

The eye (Fig. 9-2) is a fluid-filled globular structure about 25 mm in diameter that is bounded by three cellular layers. An outer protective layer (the cornea and sclera), a middle vascular layer (the choroid, ciliary body, and iris), and an inner neuronal layer (the retina).

The spaces between the cornea and lens (the anterior chamber) and between the lens and iris (the posterior chamber) are filled with a liquid called the *aqueous humor*. The much larger space between the lens and retina is filled with a jelly-like substance called the *vitreous humor*.

Light entering the eye passes through the cornea, anterior chamber, lens, and vitreous before reaching the retina. All of these must be transparent and free from optical defects if a proper image is to be formed. After a brief description of each of these structures, formation of an image by the eye is discussed.

The Cornea

The cornea is a thin (0.5 mm), transparent, avascular tissue covering the anterior portion of the eye. It is richly supplied with sensory receptors that serve to initiate protective reflexes and account for the severe pain that accompanies even minor irritation of the cornea. Oxygen is supplied to the cornea from the atmosphere, nutrients, from the aqueous humor.

The major thickness of the cornea (called the *stroma*) is composed of about 200 lamellae (layers) of tightly packed, thin fibers. If the regular spacing of these fibers is disturbed, as can occur in glaucoma (an abnormal increase in intraocular pressure), transparency is lost.

Apart from its protective function, the cornea is extremely important in image formation by the eye. The cornea has a high refractive index (e.g., is able to bend light strongly) and provides about

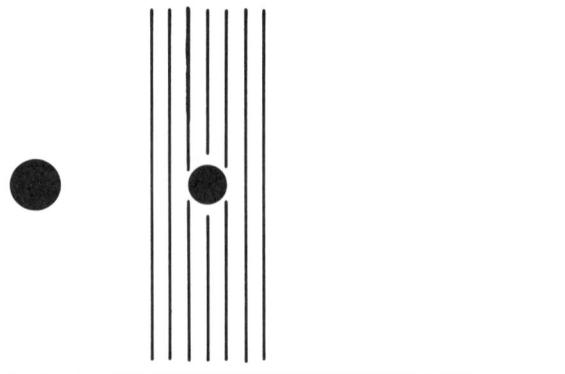

Fig. 9-1. To demonstrate the presence of a blind spot, close your left eye and look at the spot on the left with your right eye. Move the book back and forth. When the book is about 4 inches from your eyes, the spot on the right should disappear because its image is formed on the optic disk. The spot does not remain as a blank in your visual field. Instead, the cortex fills in the space with the vertical lines that occupy the rest of the visual field.

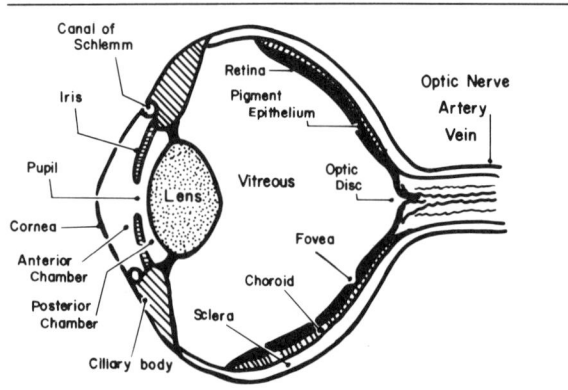

Fig. 9-2. Components of the eye required for normal vision.

two thirds of the image-forming power of the eyes.

The Aqueous Humor

The aqueous humor is actively secreted by the ciliary body into the posterior chamber. It flows through the pupil into the anterior chamber and returns to the bloodstream through the canal of Schlemm, found at the border of the cornea and sclera. About 2 microliters of aqueous humor are formed each minute. This represents about 1% of the anterior and posterior chamber volume, so that total replacement occurs in about 2 hours. As indicated above, the aqueous humor is the primary vehicle for exchange of nutrients and metabolic waste products for the cornea and lens.

The pressure exerted by the flowing fluid is normally between 10 and 20 mm Hg. This pressure is important in maintaining the shape of the eye and keeping all the refractive surfaces of the eye in proper orientation to each other. Intraocular pressure varies normally with respiration, blood pressure, and time of day. However, if the intraocular pressure rise is large and prolonged (glaucoma), damage to the cornea and retina can occur. Glaucoma usually results from a blockage of the outflow tract for the aqueous humor. It can be treated by administration of cholinergic drugs that decrease the size of the pupil and open the outflow tract or with adrenergic-type drugs that inhibit the formation of aqueous humor as well as increase outflow. If medical treatment fails to control the intraocular pressure, then surgery must be performed to prevent blindness.

The Lens

The lens is an elliptically shaped organ, about 4 mm thick and about 8 to 10 mm wide, that is completely surrounded by an elastic capsule. As the next section describes, when the eye accommodates (i.e., changes its refractive power) for near vision, the lens capsule retracts, increasing the curvature of the lens. This makes it possible to keep an image in focus as the object is moved closer to the eye.

Like the cornea, the lens is avascular and obtains its nutrients from the aqueous and vitreous humor. Transparency is possible because the lens fibers are thin and closely packed. With age, the lens can lose its transparency, forming a cataract. Cataracts can also be caused by trauma, disease, metabolic deficiency, or radiation. If the cataract interferes with vision, the lens can be removed surgically.

The Vitreous Humor

The vitreous is a gelatinous mass consisting of a diffuse network of collagen fibers and water. Its major function is to provide structural support for the retina. With age, the vitreous gel breaks down, forming pockets of water that produce the illusion of flashing lights in the visual field. Opacities in the vitreous, called *floaters*, are seen as shadows or spots. These are fairly common symptoms of age, particularly in myopic (nearsighted) individuals, and are usually harmless.

IMAGE FORMATION

Light can be described either as an electromagnetic wave or as a stream of particles, called *photons*. The model chosen depends on which best explains the property of light being discussed. For example, when light passes through a medium containing small particles, it is scattered in all directions. The amount of scatter is inversely proportional to the wavelength of light. Thus, when sunlight enters the earth's atmosphere, blue light (which has a shorter wavelength) is scattered more than green or red light (which have longer wavelengths), making the sky blue. A similar phenomenon occurs in the iris, explaining why eyes without pigment are blue. If a small amount of melanin is present in the iris, a slight bit of yellow color is added to the blue, making the eyes appear green. These phenomena are best explained by the wave-like characteristics of light.

On the other hand, it is more convenient to consider light as a stream of particles when describing the formation of an image by a lens. In this scheme, lines, or rays, of light are drawn to indicate the direction of propagation. How these rays behave when they pass through a lens to form an image is the subject of geometric optics.

Each point on an object emits light rays that diverge from each other in all directions. In order for a lens to produce an image of the object, it must bend each of the rays so that they all converge at a single point behind the lens. The path traversed by a light ray as it passes through a lens can be described geometrically and depends on two properties of the lens, its curvature and its refractive index. Lenses with a high degree of curvature (those having a small radius of curvature) bend light rays more than those with a smaller degree of curvature. Similarly, those lenses made of material with a high refractive index bend light more than those with a low refractive index.

Figure 9-3 illustrates the path followed by rays of light from a distant object as they pass through a lens to form an image. The lens is shown as part of a sphere having a refractive index, n, and a center of curvature, c. The line drawn perpendicular to the lens through the center of curvature is called the *principal axis*. Although the rays of light are actually diverging, those passing through the lens are drawn parallel to each other, because the amount of divergence is so small that it can be neglected.

Refractive Power

The refractive power of a lens is an index of how well it bends light rays. It is normally measured by determining the distance between the lens and the point at which an image of a faraway object is formed. This is illustrated in Figure 9-3. After passing through the lens, the light rays from each point on the object converge to form an image. The plane on which the image is formed is called the *image plane*, and the distance from the lens to the image is called the *image distance*.

When the object is far enough away so that the rays of light passing through the lens from each point on the object can be considered parallel to each other, the image plane is called the *focal plane*. The point at which the principal axis intersects the focal plane is called the *focal point*, and the distance between the focal point and the lens is called the *focal distance*. The refractive power is defined as the reciprocal of the focal distance. That is, $P = 1/f$, where P = power in diopters and f = focal distance in meters.

The power of the lens depends on the curvature and refractive index of the medium from which the lens is formed. Thus, in addition to being able to determine the power by finding the focal distance, the power of a lens can be calculated directly using the formula $P = (n - n')/r$, where n = the refractive index of the lens, n' = the refractive index of the medium through which the light traveled before entering the lens, and r = the radius of curvature of the lens.

From a practical point of view, rays of light emanating from objects greater than 6 m (20 ft)

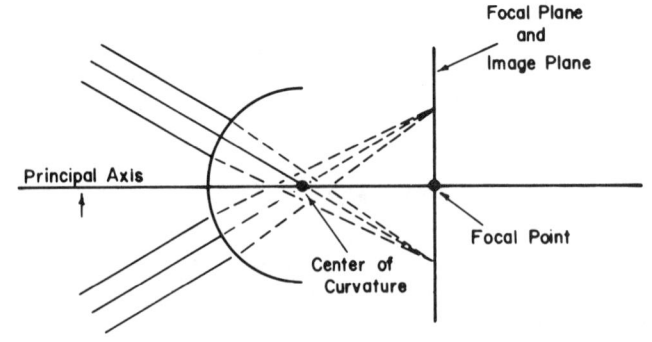

Fig. 9-3. *The path followed by light rays from a distant object.*

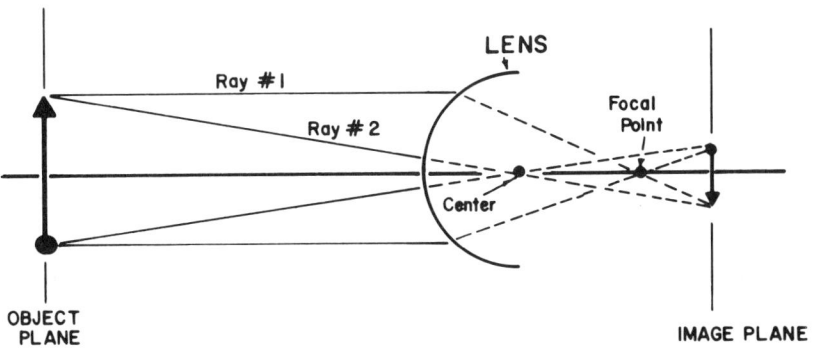

Fig. 9-4. How to find the image distance geometrically.

from the eye have diverged so little by the time they enter the eye that they can be considered parallel to each other. As a result, the image of the object is formed on the focal plane. This is why an ophthalmologist places a patient 20 feet from an eye chart when testing visual acuity.

Image Formation

Figure 9-4 illustrates the effect of moving the object closer to the lens. Now the rays entering the lens are diverging too much to be considered parallel, and the image distance becomes greater than the focal distance. The object distance, image distance, and the focal distance are related by the formula:

$$\frac{1}{o} + \frac{1}{i} = \frac{1}{f}$$

where o = the distance of the object from the lens in meters, i = the distance of the image from the lens in meters, and f = the focal distance in meters.

Figure 9-4 also illustrates how the image plane can be found graphically. One ray of light (ray No. 1) emanating from a point on the object is drawn through the center of the lens. A second ray of light (No. 2) emanating from the same point is drawn parallel to the principal axis. This second ray is bent by the lens so that it passes through the focal point. An image is formed at the point where these two rays meet. Of course, all other rays of light originating from the same point on the object also converge at this point after passing through the lens.

Image Formation by the Eye

Image formation by the eye is the same in principle as the image formation just described, but is much more complicated because the eye has a number of refractive surfaces and the image is formed in a fluid medium rather than in air. However, ophthalmologists simplify the situation by considering the eye as having a single refractive surface with a radius of curvature of 0.0068 meters and a refractive index of 1.39 (the average of the cornea and lens). Using these values in the formula given earlier, the power of the eye is calculated to be P = (1.39 − 1)/.0068 = 57 diopters. Knowing the refractive power of the eye, the focal distance can be calculated using the formula given earlier, P = 1/f or f = 1/P. Substituting a power of 57 diopters into this formula yields a focal length of 0.017 m or 17 mm. This means that a distant object (one greater than 6 m from the eye) forms an image 17 mm behind the cornea.

The retina is about 25 mm behind the cornea. The difference between this value and the one calculated results from applying the lens formula for images formed in air when calculating the focal distance. Since the errors from this simplification are generally small, the anatomical inaccuracy is preferable to applying the more complex lens formulas to ophthalmologic problems.

About two thirds of the eye's refractive power is provided by the cornea and about one third by the lens. The greater power of the cornea is a consequence of the much larger difference between the refractive power of air and cornea than between aqueous humor and lens. These differences explain why you do not see well when you open your eyes underwater; because the refractive index of water and cornea are similar, the power of the cornea is greatly reduced. Vision can be improved by wearing goggles and thus maintaining the air-cornea interface.

Accommodation

What happens if the object is moved closer to the lens? As previously indicated, this causes the image plane to be formed behind the focal plane; in the eye, the image plane is formed behind the retina. Thus, the image on the retina is out of focus. The image can be made clear again by increasing the refractive power of the lens. This is accomplished by increasing the curvature of the lens by the process of accommodation.

Normally, when the eye is relaxed, the elastic ligaments suspending the eye keep it stretched out. When the ciliary muscles are contracted, tension on the suspensory ligaments is reduced and the elastic capsule covering the lens retracts, causing the eye to bulge out. This is called *accommodation for near vision*.

For example, if the object is placed 10 cm from the eye, the image distance is

$$\frac{1}{0.100} + \frac{1}{i} = \frac{1}{0.017}$$

where $i = 0.020$ m $= 20$ mm.

Thus, the image would form three mm behind the retina and be out of focus. This is prevented by the process of accommodation. To calculate by how much the power of the lens would have

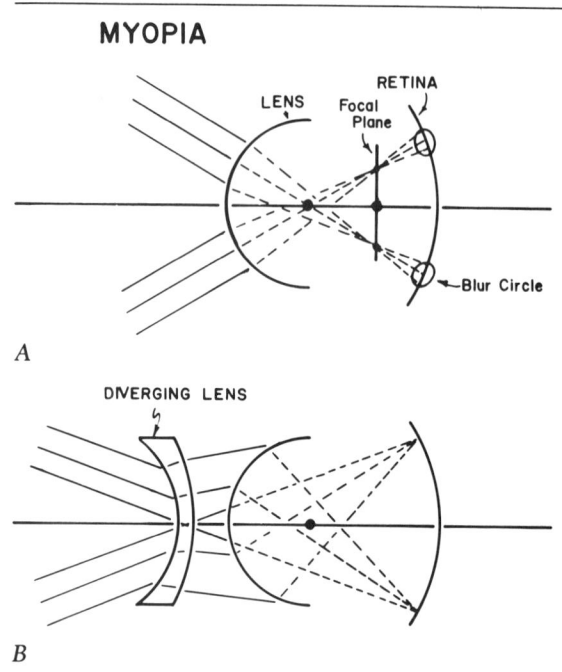

Fig. 9-5. A. *Image formation in a myopic eye.* B. *Image formation on the retina when a diverging lens of the appropriate refractive power is placed in front of the eye.*

to increase to keep the image on the retina, the lens formula is again applied. This time, however, the image distance is fixed at 17 mm and the focal length is calculated.

$$\frac{1}{0.100} + \frac{1}{0.017} = \frac{1}{f}$$

$f = 0.0145$ m $= 14.5$ mm.

This means that the focal length of the eye is reduced to 14.5 mm or the refractive power of the eye is increased to 68 diopters. That is, to keep the image on the retina when the object is placed 10 cm from the eye, the lens has to increase the refractive power of the eye by 10 diopters. The same result would have been obtained if the more complex lens formulas were used.

The maximum power of accommodation is about 14 diopters in childhood and decreases with age as the lens capsule loses its elasticity. A child can focus an object as close as 7 cm from his or her eye. This distance, the closest an object can be to the eye and still remain in focus, is called the *near point*. By age 30, accommodative power is reduced by half (7 diopters) and the near point is increased to 14 cm. This loss of accommodation with age is called *presbyopia*.

An individual who is able to see a distant object clearly without accommodation is said to be *emmetropic* (having normal vision). *Ametropia* exists when distant objects are not seen clearly. If the image in a relaxed eye forms in front of the retina, then myopia exists; if it falls behind the retina, then hyperopia exists.

Myopia

The most familiar of the refractive errors is *myopia* (nearsightedness), which occurs when the axial length of the eye is greater than the unaccommodated focal length. As a result, the image is formed in front of the retina and appears blurred (Fig. 9-5A). Myopia can result either because the refractive power of the eye is greater than normal or because the length of the eyeball is too great. If the object is moved closer to the eye, the image falls behind the focal point and eventually a distance is reached at which the image is formed on the retina. This is called the *far point* of the eye. It is the farthest an object can be from the eye and still be clearly seen. If the object is moved closer to the eye, accommodation occurs to maintain a sharp image.

Since the problem is caused by a refractive power that is too great, it can be corrected with a spectacle lens that reduces the total refractive power of the eye. As Figure 9-5B shows, this is a diverging (or minus) lens. With such a lens in front of the eyes, the focal length becomes equal to the axial length and vision of distant objects becomes clear.

Hyperopia

Hyperopia (farsightedness) is a result of having an eyeball that is too short for the refractive power of the unaccommodated eye. In this case, the actual image would be produced behind the retina (Fig. 9-6A). When the light rays from an object reach the retina they have not yet converged to a point, and thus the image appears blurred. The image can be brought into focus (onto the retina) by accommodation, but in doing so some power of accommodation is used up. Thus, the near point (achieved when accommodation is maximum) is farther from the eye in a farsighted person than in a nearsighted one. A farsighted person is not able to see distant objects with more clarity than an individual with normal vision. In fact, because some degree of accommodation is always required, even to see distant objects, eye strain is a common accompaniment of hyperopia.

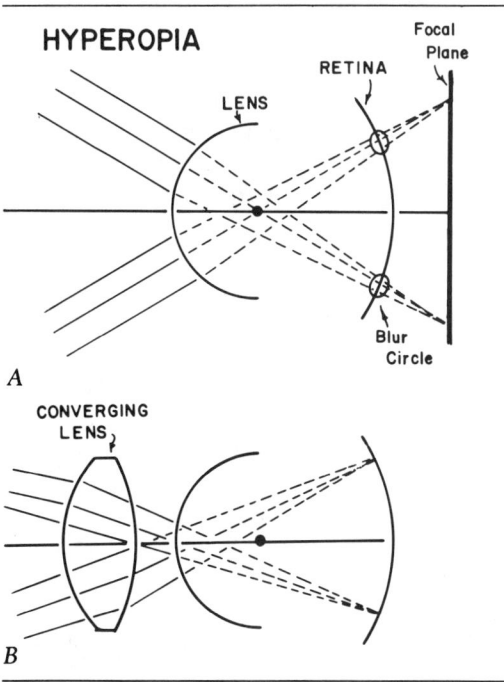

Fig. 9-6. Image formation in a hyperopic eye. B. A clear image can be formed if the appropriate converging lens is placed in front of the eye.

Hyperopia can be corrected, as shown in Figure 9-6B, by placing a converging (or plus) lens in front of the eye. Generally, if the degree of hyperopia is mild, no glasses are required because sharp vision can be obtained by accommodation. However, as the individual ages and the lens loses its elasticity, glasses become necessary. In presbyopia, a converging lens is used as a substitute for the loss of accommodative power.

At birth the eyeball is very short, producing a high degree of hyperopia. The eyeball grows rapidly during the first three years, largely eliminating the hyperopia, and then grows more slowly until puberty. It is during this latter growth stage that myopia or hyperopia develops in children. The number of individuals with ametropia requiring correction is divided approximately equally between hyperopia and myopia. About 40% of of the adult population is emmetropic, while another 35% has a small degree of hyperopia that does not require correction.

THE RETINA

The initial stages of visual processing occur within the retina. This is a well ordered, multilayered structure consisting of receptors, neurons, and supporting cells, which is divided into distinct layers. The photoreceptors are contained in the outermost layer of the retina, in terms of distance from the central nervous system. Anatomically, however, they are found at the back of the eye bordering on the pigment epithelial layer of the choroid. Next in line are the bipolar cells, followed by the ganglion cells. Horizontal cells connect neighboring photoreceptors and bipolar cells to each other in the outer layers of the retina, while amacrine cells perform a similar function for bipolar and ganglion cells in the inner layers. The supporting (glial) cells of the retina are called Müller cells. They occupy all the space not taken by the photoreceptors and neurons, and spread out to form a boundary between the ganglion cells and vitreous on one end of the retina and between the receptors and pigment epithelium on the other.

The Pigment Epithelium

The *pigment epithelium* plays two important roles in visual function. It permits the exchange of nutrients and waste products between the retina and the choroidal capillaries and is involved in the life cycle and transducer function of the rods and cones.

As there is only a loose connection between the retina and pigment epithelium, the two can become separated. Such separation usually results from age-related degenerative changes within the eye that produce small holes in the retina. Vitreous fluid seeps through the holes into the space between the retina and pigment epithelium. Once inside this space, the fluid spreads laterally, increasing the size of the detachment. This is a serious condition and surgical repair, which involves sealing the holes and reattaching the retina, must be carried out to prevent blindness from occurring.

The Photoreceptors

The retina contains about 120,000,000 rods and some 6,000,000 cones. Rods (100 microns long and 5 microns wide) are longer and wider than cones (50 to 75 microns long and 2 to 5 microns wide) (see Fig. 9-7). The rods contain a single type of photopigment and are primarily involved in detecting low levels of light. Cones are not as sensitive to light as rods, but are responsible for encoding the sensory information required for our high visual acuity and color vision.

Morphologically, the rods and cones are divided into four histological regions (Fig. 9-7): (1) an outer segment consisting of membranous disks containing the photosensitive visual pigments, (2) an inner segment containing mitochondria, glycogen granules, and protein-synthesizing organelles, (3) a cell body and nucleus, and (4) a synaptic terminal. The inner and outer segments are connected by a thin ciliated region through which nutrient and fluid exchange takes place.

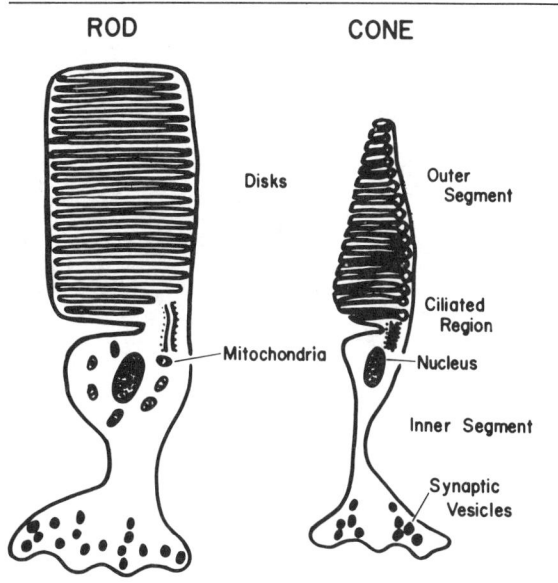

Fig. 9-7. Rod and cone. The photopigment responsible for absorbing the light rays is located within the membranes of the outer segment disks.

There are about 1700 membranous disks tightly packed in the rod outer segment. These are formed by invagination of the plasma membrane near the ciliated region. The disks are pinched off and migrate outward as new disks are formed below them. When they reach the end of the outer segment, the disks are shed and phagocytized by the pigment epithelium. It takes about 10 days for a newly formed disk to reach the top of the rod and be shed. The renewal process is very important for normal function of the retina. If the pigment epithelium is unable to phagocytize the disks, then the rods eventually die.

In cones, disks do not separate from the plasma membrane but remain as invaginations within the outer segment. There is some phagocytosis of disks, but the regular disk renewal that occurs in rods is not seen in cones.

The Photochemical Response

Rhodopsin is composed of a protein called opsin and a light-absorbing component called *11-cis* retinal (the chromophore). Rhodopsin is incorporated into the disks as they are formed, and floats in the lipid bilayer of the disk membrane with its light-absorbing end exposed to the space between the disks (Fig. 9-7). There are over one billion photopigment molecules in each rod.

The wavelength of light that is preferentially absorbed by the photopigment is a function of the molecular structure of opsin. All rods contain a single type of protein that absorbs light in the blue–green region of the spectrum. Cones contain one of three types of opsin that absorb light maximally in the blue, green, or red region of the spectrum. Since there is no universally used terminology to name these various photopigments, we will use the generic name rhodopsin to refer to the pigments of both rods and cones.

As a whole, the retina absorbs light of blue to green wavelength best, and so has a purple or magenta color. When the retina reacts with light, it loses its color and is said to be "bleached." The photochemical reaction that takes place within the rods and cones causes a change in their membrane potential.

In the dark, the chromophore of rhodopsin is found in the 11-cis conformation (Fig. 9-8). When light is absorbed by the photopigment, it initiates a conformational change in the structure of 11-cis retinal. After passing through a number of intermediate forms, 11-cis retinal is converted to the all-*trans* form. During this transformation, calcium ions are released from the membranous disks and play a role in the electrophysiologic response to light.

The Visual Cycle

Once in the all-trans form, retinal separates from opsin. It is reconverted to the 11-cis form by the enzyme isomerase, and then spontaneously recombines with opsin to reconstitute rhodopsin. At any given light level an equilibrium is reached between the amount of rhodopsin being bleached by the light and the amount being reformed enzymatically. Since the sensitivity of the eye to

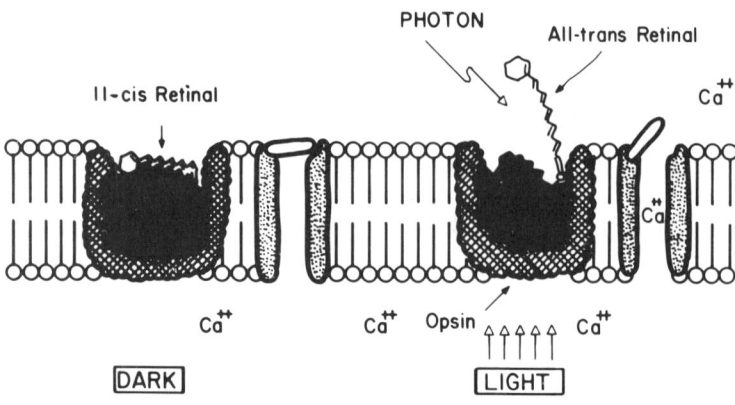

Fig. 9-8. Response of the photopigments to light.

light depends on the amount of unbleached rhodopsin, these competing processes permit the sensitivity of the eye to change with changing levels of illumination.

The retinal that is not reconverted immediately to the 11-cis form is reduced to retinol (vitamin A). The retinol then diffuses out of the cell and into the pigment epithelium, where it is stored. If the light level is reduced, the retinol is oxidized to retinal and returned to the receptor cell, where it can be reattached to opsin. Retinal is continuously lost from the eye by normal degradative processes. It is normally replaced with retinol (vitamin A) removed from the blood by the pigment epithelium. Night blindness (loss of visual sensitivity at night) is a common result of vitamin A deficiency, but can also be caused by disruption of the normal activity of the pigment epithelium.

The Receptor Potential

Vertebrate rods and cones differ from all other receptors in that they respond to stimulation by hyperpolarizing rather than by depolarizing. Prior to excitation, the photoreceptors have a low resting potential of about -40 to -50 mV, due to a high resting permeability to sodium. This causes a fairly high flow of sodium ions into the outer segment. The sodium passes through the ciliated region into the inner segment, where it is pumped out of the cell. When stimulated by light, the permeability to sodium decreases and the membrane hyperpolarizes.

The mechanism by which light causes the decrease in sodium permeability is not known, but based on the available evidence, two theories have been proposed. One suggests that sodium channels in the outer segment are controlled by the cytoplasmic calcium concentration; as calcium levels rise, sodium channels close, causing the membrane to hyperpolarize. As indicated earlier, one of the consequences of the photoisomerization of 11-cis retinal is the release of calcium from the outer segment disks. It is this calcium that is thought to cause the inactivation of sodium channels. The response is terminated by the resequestering of calcium into the disks by membrane-bound pumps.

The other theory proposes that the substance controlling the activation of the sodium channels is cyclic guanosine monophosphate (c-GMP). According to this concept, c-GMP activates a protein kinase that phosphorylates the sodium channel, keeping it in the open state. Thus, the membrane remains depolarized when the rods and cones are not exposed to light. When rhodopsin absorbs light it activates a protein, called transducin, which in turn causes the activation of a membrane-bound phosphodiesterase. The phosphodiesterase then hydrolyzes the c-GMP, causing the sodium channels to close and the membrane to hyperpolarize.

Although the exact mechanism by which phototransduction occurs is not yet known, it is clear that the only function of light is to initiate the photoisomerization of 11-cis retinal to its all-trans form. All other events in the phototransduction process then proceed spontaneously.

The hyperpolarization of the receptors is the initial step in the transfer of information to the visual cortex. The encoding of this information within the retina is complicated and is not yet completely understood. However, what is known about its neuronal processing helps explain a great deal of the perceptual capabilities of the visual system.

Sensory Processing Within the Retina

The neuronal circuitry of the retina is organized in both vertical and horizontal fashions (Fig. 9-9). Vertically, rods and cones synapse with bipolar cells, which in turn synapse with ganglion cells. Laterally, horizontal cells form synaptic contacts between neighboring bipolar cells, while amacrine cells connect ganglion cells together.

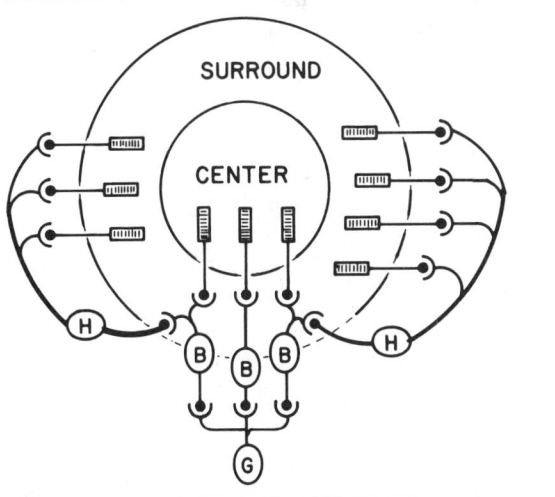

Fig. 9-9. Cellular organization of the retina.

The On-Center Receptive Field

The synaptic organization can best be understood with reference to the ganglion cell's receptive field. Figure 9-9 illustrates the inputs to a single ganglion cell. The ganglion cell receives an input from a number of bipolar cells. Each of these bipolar cells, in turn, receives an input from a large number of receptor cells. All of these receptor cells are contained in a circular area on the retina that varies from 0.1 to about 2.0 mm in diameter, depending on its location. Ganglion cells within the fovea have the smallest receptive fields, while those in the more peripheral regions of the retina have larger receptive fields. When light strikes any of these receptors, it causes the ganglion cell to increase its rate of firing.

The Off-Surround Receptive Field

Another group of receptor cells is also able to affect the firing rate of the ganglion cell. These cells are contained in an annular ring surrounding the circular region. They are connected to the bipolar cells through the horizontal cells. When light strikes any of these receptors, it causes the firing rate of the ganglion cell to decrease. This ganglion cell is thus described as having an on-center, off-surround receptive field, because it is turned on when light strikes the center of its field and turned off when light strikes the receptors surrounding the center.

Prior to being stimulated, when sodium permeability is high and the membrane is depolarized, the photoreceptors are continually releasing neurotransmitter. The identity of the transmitter is not known, but it is assumed to be inhibitory because it keeps the bipolar cell in a hyperpolarized state. This, of course, means that the bipolar cell cannot release its neurotransmitter, and thus the ganglion cell is not stimulated.

When light strikes the receptors, the membrane hyperpolarizes, and the amount of neurotransmitter being released is reduced. Since the receptor's neurotransmitter is inhibitory, this reduction in neurotransmitter release results in the bipolar cell being released from inhibition. The bipolar cell thus depolarizes, releases an excitatory transmitter, and causes the ganglion cell to fire. This accounts for the on-center nature of the receptive field.

The off-surround property is mediated by the horizontal cells. When light activates (hyperpolarizes) the receptors in the receptive field surround, the effect is to stimulate the horizontal cells. These cells then release an inhibitory neurotransmitter that hyperpolarizes the bipolar cells in the field center, causing a reduction in their transmitter release and an inhibition of the ganglion cell.

The Off-Center, On-Surround Receptive Field

In addition to the on-center, off-surround ganglion cells there are also off-center, on-surround ganglion cells that behave, as their name implies, in an opposite way. That is, they are inhibited when light strikes the center of their field and excited when light strikes their surround. They most likely respond in this fashion because their bipolar cells respond in the opposite way to the transmitter released by the photoreceptors. These cells, unlike the on-center bipolar cells, are normally excited by the transmitter released by the receptors. Thus, when light strikes the eye and the amount of transmitter released is decreased, the off-center bipolar cells become hyperpolarized. As a result, they release less transmitter and the ganglion cell firing rate is decreased. Quite possibly, a single transmitter released from a cone activates one bipolar cell (in an on-center field) and inhibits another (in an off-center field).

Whether these two types of pathways play fundamentally different roles in information processing or simply provide for redundancy is not known. This discussion considers only the on-center, off-surround receptive field.

All of the neuronal processing in the retina discussed so far is accomplished without action potentials. This is possible because the cells are very small and information can be transferred by the passive spread of current. The ganglion cells, of course, must produce action potentials to transmit their information into the central nervous system. The major consequence of this type of neuronal organization is that a diffuse light stimulus that affects the entire receptive field (center and surround) does not activate the ganglion cell and thus is not transmitted to the central nervous system. On the other hand, stimuli that activate only the center of the field cause the ganglion cell to fire, while stimuli directed to the surround of the field cause the ganglion cell to stop firing. This type of response to light enhances the acuity of the visual system (see Visual Acuity).

The Geniculate Nucleus

Ganglion cells from both eyes project to the lateral geniculate nucleus in an orderly fashion. Histological studies reveal that the geniculate nucleus is divided into six layers. Optic tract fibers (axons of the ganglion cells) from the ipsilateral eye terminate on layers two, three, and five, while those from the contralateral eye end on layers one, four, and six. The functional value of segregating the retinal input in this manner is not known.

The receptive field properties of the retina are preserved in the geniculate neurons. Although it is likely that some neuronal processing does occur within the geniculate nucleus, it appears that its major function is to relay information from the retina to the occipital cortex.

THE VISUAL CORTEX

The visual cortex receives its input from the *geniculate nucleus*, which in turn is innervated by the *retinal ganglion cells*. There are three types of retinal ganglion cells, X, Y, and W cells. The axons of all three types of retinal neurons leave the eye through the optic disk, where they join together to become the optic nerve. The X ganglion cells have the smallest receptive fields within the retina and project to a relatively small area of the geniculate nucleus. This precise representation of the retina is maintained within the cortex. That is, the fovea, which is capable of producing the greatest degree of visual acuity, has the greatest amount of cortical area devoted to it, while extrafoveal regions of the retina, which cannot produce sharp images, project to much smaller areas of the cortex. This retinotopic organization is analogous to the somatotopic organization of the cutaneous sensory system described previously in Chapter 7, Cutaneous Sensation.

Although some of the Y ganglion cells project to the geniculate, most of them and all of the W cells project to the superior colliculus. In addition to receiving information from the retina, the superior colliculus also receives input from the visual, sensory, and auditory cortexes and thus is able to coordinate the head and eye movements necessary to direct the eyes to the source of an external sensory stimulus. Our discussion is limited to the X ganglion cells, which are responsible for producing the high degree of visual acuity, color vision, and depth perception that we are capable of achieving.

The geniculate nucleus receives input from both eyes. Fibers from the nasal portion of the left eye and the temporal portion of the right eye pass to the right geniculate nucleus. Thus the right geniculate nucleus receives its input from the left visual field. Similarly, the left geniculate nucleus receives its input from the nasal portion of the right eye and the temporal portion of the left eye, and so is responsible for transmitting information about the right visual field.

The axons from the geniculate nucleus, called the *optic radiation*, project ipsilaterally to the primary visual cortex, located in Brodmann's area 17 of the occipital lobe. The segregation of input from the left and right eyes is maintained within the first group of cells receiving input from the geniculate nucleus; groups of cortical cells receiving input from the left eye alternate with groups of cells receiving input from the right eye. However, cells receiving input from the left and right eyes project onto the same neighboring cells. These cells, receiving input from both eyes are responsible for generating our perception of depth. Recall that the left visual cortex, although receiving input from both eyes, receives information about the right visual field and that the left visual cortex receives its information from the right visual field. However, despite the precise retinotopic representation that is formed on the visual cortex, visual perceptions are not created by reproducing the retinal image on the cortex. Instead, they are formed by integrating information from many cortical cells, which respond in unique ways to a sensory stimulus.

Attributes a stimulus needs to activate a cortical neuron are revealed by studying their receptive field organization. Knowing how these neurons respond to retinal stimulation explains the neurophysiological basis for our perceptual capabilities.

Receptive Field Organization of the Cortex
Three types of cortical receptive fields have been described, and are classified as simple, complex, and hypercomplex. Simple units have receptive fields that are organized in a linear fashion (Fig. 9-10). When the central portion of the receptive field is illuminated, the neuron fires. Inhibition is caused when the light is projected to the lateral borders of the receptive field. Other receptive fields are stimulated when light is projected to the lateral regions and inhibited when it is projected to the central region. Still others have excitatory and inhibitory areas lying side by side. Moreover, to be effective the stimulus has to be oriented at a particular angle. Some cells only respond if the stimulus is vertical, while others may require a horizontal orientation. Still others require that the bar of light be oriented at some oblique angle.

Complex cells are similar to the simple cells in that they respond to properly oriented stimuli. But unlike the simple cells, they do not have an excitatory-inhibitory organization, so stimuli anywhere within their receptive fields causes excitation. In addition, complex cells respond best when the stimulus is moving within their receptive field.

Hypercomplex cells have, as their name suggests, relatively complex receptive fields. Like the complex cells, stimuli must be oriented in a particular direction and move within the visual field in order to elicit a response. However, they differ from the complex cells in that the stimulus must be of a particular shape. A bar of light stimulating the cell might have to be of a particular size. If it becomes longer or shorter, its effectiveness decreases. Other hypercomplex cells only respond to stimuli shaped like a corner or a triangle.

Cortical Columns

All of these cells are contained within sensory columns extending from the cortical surface to the underlying white matter. Each area on the retina is represented by a rectangular hypercolumn approximately 1 mm wide by 2 mm long. Within each hypercolumn there is a set of 18 columns, each of which contains neurons with simple receptive fields that respond to a bar of light oriented at a particular angle. For example, if the simple column at the edge of a hypercolumn contained neurons that responded to a vertically oriented bar of light, the next simple column would contain cells that responded best to a bar of light oriented at a 10° angle to the vertical, and so on until all 360° of possible orientations had been encoded. Within each simple column, the neurons receiving input from the left eye are segregated from those receiving input from the right eye. In addition, there are neurons within each column receiving bilateral inputs. Also within each hypercolumn there is a group of cells, organized in a cylindrical or peg-shaped manner, that are interposed between the simple orientation columns. These peg-shaped columns are thought to receive information from cones and to be involved in the interpretation of color. The neurons within area 17 project to neurons within areas 18 and 19 on the visual cortex, where an even more complex receptive field organization is found.

Although the precise mechanism by which the columns operate is not understood, their organi-

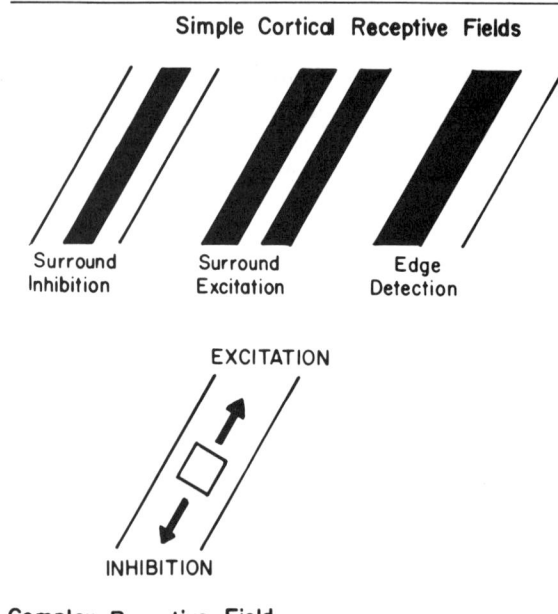

Fig. 9-10. *The receptive field of simple cortical cells is rectangular. Light must be oriented at a particular angle if it is to affect the firing rate of the cell.*

zation supports the notion that visual perceptions do not emerge from the retinotopic projection of an image onto the visual cortex. Instead, each area of the retina sends information about many aspects of a visual scene (such as its intensity, contrast, shape, motion, and color), all of which are processed in parallel to produce a coherent perception. The next section discusses how the receptive field properties of the neurons within the visual system explain our perceptual capabilities.

VISUAL PERCEPTION

The previous sections describe the peripheral and central mechanisms of information processing used by the visual system. This section considers the perceptual capabilities of the visual system and explains the neurophysiological mechanisms by which they are achieved.

Visual Acuity

Visual acuity is a measure of how well the visual system is able to distinguish two light stimuli separated in space (i.e., spatial resolution). Visual acuity can be assessed in a variety of ways, the most common of which is to use the familiar Snellen eye chart. Use of the chart is explained in the next section on assessment of visual function. For now it is only necessary to point out that visual acuity is highest when the image is focused on the fovea, the illumination of the object is bright, and there is a high contrast between the letters and the background of the chart.

Visual acuity is highest when the image is formed on the fovea (the central region of the retina about 1 mm in diameter). This phenomenon is related to a number of factors. First, the image is clearest on the fovea because the receptors are not covered by as many cellular elements as they are in other regions of the retina. Recall that for light to reach the outer cell layer it must pass through the blood vessels that lie on the surface of the retina and all the neuronal layers within the retina. Blood vessels do not travel over the fovea and the ganglion cell axons are diverted from the fovea so light can reach the receptors with a minimum of distortion. Second, the cones within the fovea are the smallest and the most densely packed in the retina, and have the least amount of convergence onto ganglion cells. In order for two light stimuli to be seen as separate images, they must fall on two cones that are separated by an unstimulated cone. The small size and high concentration of the foveal cones make this possible with a minimal spacing between stimuli. To take full advantage of their small size, each cone must project to a different ganglion cell. In the fovea there are a group of ganglion cells, called *midget cells*, that receive input from only one receptor, while others have small receptive fields. This lack of convergence contributes to the high visual acuity of the fovea. Images formed away from the fovea are not seen as well because the extrafoveal ganglion cells have much larger receptive fields. This is analogous to the situation on the skin, where small receptive fields enhance two-point discrimination.

Another factor that influences visual acuity is the amount of light stimulating the retina. If light levels are reduced, then accommodative reflex causes the pupil to increase in size and admit more light. The larger aperture of the pupil causes the optical defects of the eye to be exaggerated, and thus reduces the image-forming capability of the eye. Of more importance, however, is the effect of light intensity on receptive field size. When light levels are high, stimulation of horizontal cells is maximum and the size of the on-center field is reduced to a minimum, thus decreasing convergence and consequently increasing acuity. Reduction of stimulus intensity causes a corresponding decrease in lateral inhibition, increase in receptive field size, and loss of acuity. This increased convergence contributes to a heightened sensitivity to light, which is beneficial under low light conditions.

Dark Adaptation

If you enter a dark room after spending time outdoors in bright light, you notice that at first you cannot see anything, but that over time your eyes adjust to the dark and you are able to see quite well. This process is called *dark adaptation* and is primarily a function of the change in concentration of unbleached rhodopsin found in the rods and cones.

Figure 9-11 illustrates the result of an experiment carried out to study the time course of dark adaptation. A person is first exposed to a bright light that bleaches most of the rhodopsin in the eyes, and then placed in a dark room. The visual threshold is tested periodically by determining the lowest intensity of light that is perceptible. Notice that there are two components to the dark adaptation curve. The first represents the recovery of rhodopsin in the cones. This is complete in about 5 to 7 minutes. The second component is due to regeneration of rhodopsin in the rods. It takes about 10 minutes for the sensitivity of the rods to exceed that of the cones and about 30 minutes for rod sensitivity to reach a maximum. As Figure 9-11 shows, the relationship between recovery of the rhodopsin and threshold to light is logarithmic, so that about 90% of rhodopsin in rods must be regenerated before their sensitivity to light exceeds that of cones. Overall, sensitivity to light increases 1,000,000-fold during dark adaptation.

Also contributing to the greater sensitivity to light after exposure to a darkened environment is the increased amount of convergence, discussed earlier. The more receptors available to stimulate the ganglion cells, the greater the amount of spatial summation and the smaller the signal from each receptor required to activate the ganglion cell. Of course, with the increased receptive field size comes reduced acuity. Similarly, the widening of the pupils (from about 2 mm to a maximum of 8 mm) allows more light to enter the eye which, while increasing sensitivity, causes a reduction in acuity.

In the dark adapted state vision depends entirely on rods, and so no appreciation of color is possible. This situation, when only rods are used, is called *scotopic* vision. *Photopic* vision refers to the light adapted state when only cones are used, while *mesopic* vision occurs in the transition zone (during illumination by moonlight, for example), when both rods and cones are active.

Perception of Contrast

The discussion of the blind spot pointed out that perceptions are not simple reconstructions of retinal images. This is also evident in considering our perception of brightness, which depends on the contrast between an object and its background. If a white disk is projected onto a screen in a darkened room, and the intensity of the light in the surrounding annulus is progressively increased, the disk appears to become darker. This is true even though the actual amount of light reflected from it, or its luminosity, does not change. The contrast illusion can be explained using the information presented earlier about the receptive field organization of the retina.

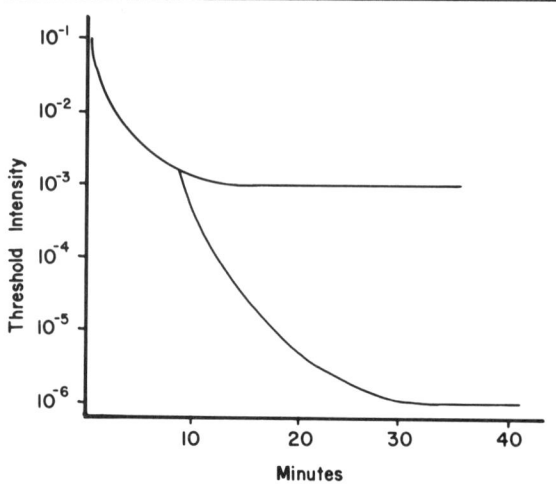

Fig. 9-11. Results of an experiment in which the ability of an individual to detect a spot of light is tested at various times after being exposed to a bright light that bleaches all of the photopigment. The cones recover their maximum sensitivity before the rods do, but do not ever become as sensitive to light as the rods.

Recall that the receptive field of each ganglion cell is organized in a center-surround fashion. The frequency of firing developed by the ganglion cell depends on the relative amount of light in the center and surround of the receptive field and not on the absolute level of illumination. Thus, if the light reaching the surround of the receptive field of the ganglion cell is much less than the light reaching the center, the ganglion cell is free from inhibition and the area included within the center of the field appears to be very bright. By increasing the amount of light reaching the surround, the inhibition of the ganglion cell is increased, and thus its firing rate is decreased. This makes the center of the field appear darker, even though there is no alteration in the amount of light reaching it.

This receptive field property also provides a basis for understanding why the cortex is able to "fill in" parts of the visual field that are missing. The ganglion cells that have receptive fields con-

Fig. 9-12. The figure appears as a rabbit when held vertically, but as a duck when held horizontally. From L. G. Brandes, Introduction to Optical Illusions. *Portland, ME: J. Weston Walch, 1976. Used with permission.*

tained entirely within a part of an image that has a constant illumination do not change their firing rate, because both the center and surround of these fields are equally illuminated. Since these ganglion cells do not change their level of activity, they are not used by the brain to reconstruct the image. In producing a perception, the cortex "assumes" that the area of the image served by those ganglion cells that are not active is of the same color and brightness as the surrounding areas where the ganglion cells are active. Moreover, because of these properties of the ganglion cell receptive fields, contrast is actually enhanced at the borders of the image where the level of illumination is changing. Recall that this occurs because the difference in activity between ganglion cells at the border between the light and dark areas is increased by lateral inhibition. These phenomena reinforce the concept that the visual processing apparatus is organized to respond to changes in illumination or color and not to constant levels of stimulation.

Pattern Recognition

Certain patterns in a retinal image are recognized easily by an observer. For example, if a page is filled with dots in a random fashion, an observer is usually able to group some of them into familiar patterns such as a circle or a polygon. Similarly, the simple line drawing illustrated in Figure 9-12 is interpreted as a rabbit by the cortex, but if the book is turned on its side, the same image is seen as a duck.

The receptive field properties of neurons within the visual cortex provide a basis for understanding why recognizable images are so easily formed by the cortex. Recall that within each orientation column there are cells specialized to respond to stimuli that have a particular angle and shape. Thus, the apices of a triangle or the edges of a figure stimulate cells within a column, while the rest of the image has little or no effect on cortical neurons. The locations of the activated cells are then used to create the boundaries of the perception and prior experience is used to fill in the missing parts.

A variety of experiments have been performed to demonstrate the necessity of having normally organized cortical receptive fields if pattern recognition is to be possible. These fields are not present at birth and require normal visual experience for their proper development.

Cats raised from birth with their eyes covered by a mask that permits viewing of only horizontal stripes do not develop cortical orientation cells responsive to vertical stimuli. These animals were found to be deficient in their ability to detect and respond to stimuli presented in a vertical direction. A similar disability develops in children who do not use one eye as infants. If the muscles of the two eyes are not equally strong, convergence does not occur and the two eyes view totally different visual fields. This causes double-vision, *diplopia*, to occur. The central nervous system responds by supressing the image from one of the eyes, resulting in the failure of cortical neurons to develop normal receptive fields. Consequently, there is loss of pattern recognition and the child actually loses sight in the affected eye. This condition, in which a disturbance of visual acuity occurs due to disuse of the eye, is called *amblyopia*. To prevent it from occurring, ophthalmologists generally place a patch over the preferred eye for part of the day to ensure that both eyes are used. This should be done before the age of six or seven, because by that age the development of the visual pathways is essentially complete and abnormalities can no longer be corrected.

Color Vision

Our ability to perceive color is a result of the light-absorbing characteristics of the cone photopigments and the sensory encoding mechanisms of the visual system. As explained earlier, there are three types of cones, each one absorbing light over a different portion of the visual spectrum. They are named for the color corresponding to the peak of their light-absorbing curves. Thus, there are red, green, and blue cones, each containing a different photopigment, named erythrolabe, chlorolabe, and cyanolabe, respectively.

The color we perceive is determined by the combination of cone receptors stimulated. If the eye is stimulated by light having a wavelength of 600 nm, the red cones are predominantly stimulated and we perceive the color red. If, however, the light has a wavelength of 570 nm, then the red and green cones are equally stimulated and we perceive the color yellow. If all three cones are equally stimulated, then the color we perceive is white. In fact, we can perceive all the colors of the spectrum by varying the amount of activity generated in each cone, just as the three primary colors can be used to produce any of the spectral colors if they are mixed in the appropriate proportions.

Color Blindness

Not everyone sees colors in the same way. Individuals whose cones have pigments that absorb light over a range of the spectrum that is different from normal are said to have *anomalous color vision*. If the chlorolabe in the green cone absorbed light maximally at 540 nm instead of 535 nm, a light that is called yellow by a normal individual appears greenish-yellow to an individual with anomalous color vision. This difference in perception occurs because the yellow light (with a wavelength of 570 nm) has a greater effect on the green cone than it does on the red cone, whereas normally the yellow light affects the red and green cones equally.

If one of the pigments (red, green, or blue) is missing altogether, color blindness, or more properly, color deficiency results. The most common of these deficiencies is due to the complete absence of either the red or the green pigment. This disorder is called *dichromacy* because there are only two pigments instead of three. If the green pigment is missing it is called *deuteranopia* (absence of the second primary color) and if the red pigment is lacking, it is called *protanopia* (absence of the first primary color). Individuals with dichromacy cannot distinguish between colors in the green to red portion of the spectrum, because they have only one light-absorbing pigment in this range of wavelengths. The pigment that remains in deuteranopia, the red pigment, is found in both the red and the green cones. Thus, colors in this part of the spectrum are perceived as shades of yellow, because no matter what the wavelength of the light striking the retina, the red and the green cones are equally stimulated because they all contain the same pigment type.

Not all of our color perceptions can be explained simply as a function of the light-absorbing properties of the cones. Some are based on the information-processing characteristics of the retinal ganglion cells receiving input from the cones. The retinal ganglion cells are organized in an antagonistic or opponent fashion similar to that discussed earlier in terms of acuity and contrast. In this case, however, it is color rather than luminosity which determines the activity of the cell.

In some ganglion cells, green light may activate the cell while red light inhibits it. In others, blue and yellow have opposite effects on the ganglion cell. This behavior of the ganglion cell can be used to explain the chromatic afterimages that are seen after staring at an object of one color and then looking at a piece of white paper. If you look continuously at something red for 30 seconds and then shift your gaze to something white, the afterimage of the red object has a green color. This presumably occurs because the red-green opponent cells, which are inhibited by red light, are excited when the inhibition is removed, thus signalling to the cortex that a green stimulus has been applied. Similar receptive field properties exist in the geniculate nucleus and visual cortex.

We have been emphasizing the contribution of receptive field properties to our visual perceptions. The contribution seems especially great when considering how a flat retinal image can be perceived as having three dimensions.

Depth Perception

The world we observe is perceived in three dimensions, even though the image formed on our retina is two-dimensional. Artists have learned a variety of methods to create the illusion of depth in their two-dimensional paintings. For example, objects drawn larger than others are perceived to be closer. Similarly, objects drawn with less clarity or brightness are assumed to be farther away. Shading can also be used to create an impression of depth. None of these techniques provides a sensation of depth comparable to that created by the makers of so-called 3-D movies, however. The three-dimensional view in these movies requires binocular vision and is called *stereopsis*. The following paragraphs explain the neurophysiological basis for our stereoscopic vision.

For a moment, stop reading this book and look into the distance. You should perceive a single view of the room. However, the images formed on your two eyes are not exactly the same. You can demonstrate this by holding your finger at arm's length in front of your eyes. Close first one eye and then the other. Notice that the object that is directly beyond the finger shifts as you alternate between the left and right eyes. If you line up your finger with two objects at different distances from your eye, you will notice that the amount of apparent movement between your finger and the object is greater for the object farther from your eye. The difference between the view of the left and right eye is used by the central nervous system to create the sensation of depth.

Figure 9-13 shows the geometric basis for this disparity. Both eyes converge so that the image of object A falls on the foveae of both eyes. Objects B and C do not form images on both foveae because they are at different distances from the eye. Their images are on noncorresponding, or disparate, parts of the retina. The disparity between the images found in the two eyes could, by trigonometric methods, be used to calculate the depth of the objects relative to the fixation point. Within the visual cortex are neurons that perform an equivalent function and produce our perception of depth.

Previously, we discussed the organization of the visual cortex into columns containing cells which respond to the orientation of lines. The activity of these cells is also dependent on whether they receive input from one or both eyes. A line of the appropriate orientation thus stimulates the neuron only when it is imaged on the foveae of both eyes. When the image falls on the fovea of one eye and on a nonfoveal point on the other eye, the amount of neuronal activity is reduced. A neuron with this response is called a *feature detector* because it integrates information received from more than one neuronal input.

Other feature detectors respond best when the image falls on the fovea of one eye and a nonfoveal region of the other eye. When this neuron is activated, it indicates to the central nervous system that the object viewed by this particular column is not on the fixation point but is at some distance from it. There are neurons within the column that respond best to various degrees of image disparity. The particular neuron within the column that is activated by the image indicates how far the object is from the point of fixation. How the central nervous system actually puts together all the information encoded by these neurons to produce our visual perceptions is not at all understood. But these mechanisms can be used to explain how 3-D movies are produced.

When making the movie, the scenes are photographed using two different camera angles. When shown, the two movie projectors are covered with different colored filters and the viewers wear goggles with similarly colored lenses, so that each eye sees the scene projected by only one of the cameras. Since the two cameras viewed the scene from different angles, the images are displaced from one another *except* at the fixation point. Thus, the images formed on one of the retinas are displaced, and the disparity that results is used by the central nervous system to produce a vivid perception of depth.

We have just described the physics of image formation, explained how visual transduction occurs and how the nervous system is organized to interpret visual information, and provided a neurophysiological basis for understanding a variety of visual perceptions. This chapter concludes with a brief description of some common methods of examining visual function.

CLINICAL ASSESSMENT OF VISUAL FUNCTION

There are a great number of instruments available to the ophthalmologist for examining the visual system. The methods described here are simple

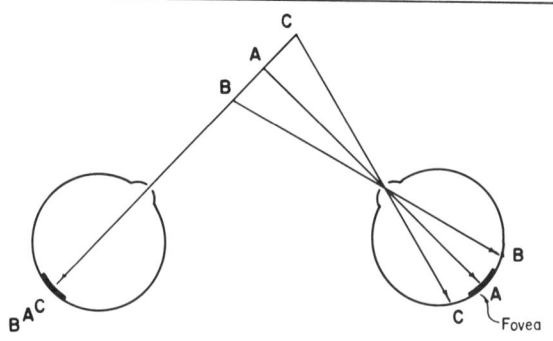

Fig. 9-13. *The geometric basis for depth perception. The image of an object fixated by both eyes (A) falls on the foveae. Objects closer or farther from the eye do not fall on corresponding points of both eyes. The image of object B falls to the left and the image of object C falls to the right of the fovea in the right eye, whereas both images fall on the fovea in the left eye. The disparity between the location of the images in the two eyes is used by the brain to produce a three-dimensional perception of flat images formed on the retina.*

techniques that are used by a family practitioner in his or her office as part of a complete medical examination.

Measuring Visual Acuity

As indicated earlier, visual acuity can be assessed quite easily using the Snellen eye chart. The letters making up the chart are drawn so that each component forms a retinal image 5 microns wide, or a visual angle of 1 minute, if placed at the indicated distance. This size is chosen because it represents the normal resolving power of the eye.

The chart is placed at a viewing distance of 20 ft (or 6 m), because at this distance accommodation for near vision is minimal for an emmetropic individual. When tested, individuals should be able to resolve the letters on the 20-foot line. If they can do so, their visual acuity is recorded as 20/20, meaning that they can resolve letters at 20 feet as well as an emmetrope can. If, however, the smallest letters they are able to recognize are on the 50 foot line, their vision is recorded as 20/50, meaning that they have to be at 20 feet to see what an emmetrope can resolve at 50 feet.

The test is done under optimal viewing conditions: The image is formed at the fovea, illumination is bright, and there is a high contrast between the letters and background of the chart. Under these conditions, the test is an accurate measure of the image-forming capability of the eye's optical system, assuming, of course, that individuals are able to recognize and name the letters on the chart.

Measuring Intraocular Pressure

Tonometry is a technique for indirectly determining intraocular pressure. There are two methods commonly employed for making this measurement. The simpler and more practical method uses an instrument that measures the amount of indentation produced in the cornea when a known weight is placed on it. A more accurate determination uses an instrument that measures the amount of force required to flatten a small area of the cornea. Since the latter instrument is more expensive and difficult to use, it is only used by ophthalmologists.

If the intraocular pressure is found to exceed 20 to 25 mm Hg, the patient is assumed to have glaucoma. Treatment is necessary to relieve the pressure in order to prevent blindness from occurring.

Viewing the Retina with an Ophthalmoscope

The ophthalmoscope is one of the most useful instruments in medicine. It permits a direct examination of the interior of the eye that can, in addition to discovering pathologies of the eye, also detect the presence of systemic diseases such as diabetes or hypertension. Its most important feature is its light source, which can be directed into the pupil of the eye without obstructing the examiner's view of the eye. In performing an ophthalmoscopic examination, the physician places his or her eye close to the patient in order to see as much of the retina as possible. Focusing on the retina is accomplished without accommodation for near vision because, from an optical point of view, the retina appears as a distant object. This is because the retina of an emmetropic individual is at the focal plane of the eye, which means that rays of light from a far-away point (that are parallel to each other) form an image on the retina. The converse is also true; rays of light emerging from the retina leave the eye parallel to each other. Thus, when these rays enter the examiner's eye they come to focus on the retina, assuming that the examiner is also emmetropic or is wearing corrective lenses. Of course, if either the patient or the examiner is ametropic or is accommodating for near vision, the retina is not seen clearly. However, these refractive errors can be compensated for using lenses built into the ophthalmoscope.

10 Hearing

Hearing, like vision, allows us to become aware of our surroundings without having to make physical contact with them. Of even more importance, hearing makes it possible for us to communicate verbally with each other and has thus played an essential part in the development of modern societies. This chapter discusses how sounds are transduced into electrical signals within the ear, how auditory information is encoded by the nervous system, and the neurophysiological basis for some of our auditory perceptual capabilities. It concludes with a brief description of the types of hearing loss and the methods used to assess them.

SOUND STIMULUS

The stimulus eliciting a sound sensation is a pressure wave that is propagated through the air to the ear. It is generated by any condition that causes air to vibrate, such as two rocks being struck together or the vibration of a tuning fork. The air particles that are set in motion by the vibrating object do not actually travel to the ear. Instead, they move back and forth, thereby producing regions of increased and decreased pressures, and causing neighboring particles of air to move along with them. Over time, particles farther and farther from the original energy source are made to vibrate, and the pressure wave eventually reaches the ear. In air, sound waves propagate at about 340 meters per second (or 720 mph).

Measurement of a Sound Stimulus

The pressure wave is characterized in terms of its frequency and intensity. Frequency is given in terms of the number of vibrations or cycles per second. The unit used to describe frequency is the hertz (Hz), defined as one cycle per second.

Human hearing operates over a range of approximately 20 to 20,000 Hz.

The Decibel

Sound intensity can be measured in terms of energy (watts/cm^2) or in terms of pressure (Newtons/m^2 or pascals). Pressure is used more often because it is easier to measure. The threshold for human hearing at 1000 Hz is approximately 2×10^{-5} N/m^2. Rather than note the actual magnitude of a sound stimulus, audiologists describe the intensity of a sound stimulus as a logarithmic function of the ratio between the sound intensity being measured and some reference intensity. The ratio is called the *sound intensity level* (IL) and the unit of measurement is the decibel (dB). The formula to calculate the ratio is

$$IL = 10 \log \frac{I}{I_0}$$

where I = the intensity of the sound measured and I_0 = the reference intensity level, which is 10^{-16} watts/cm^2.

When sound pressure instead of power is used as the unit of measurement, the ratio is called the *sound pressure level* (SPL) and the formula is

$$SPL = 20 \log \frac{P}{P_0}$$

Table 10-1. Intensity of some common sounds

Sound intensity (dB SPL)	Typical sound
0	Threshold of hearing
10	Normal breathing
20	Whisper
40	Classroom noise
60	Normal conversation
80	Traffic noise
100	Subway train
120	Discomfort
140	Jet engine
160	Pain

where P_0 is the threshold for human hearing (2×10^{-5} N/m^2). The 20 results from the conversion from units of power to pressure. Since I is proportional to P^2

$$SPL = 10 \log \left(\frac{P}{P_0}\right)^2 = 20 \log \frac{P}{P_0}$$

Thus, a sound intensity 10 times threshold is 20dB, and an intensity 100 times threshold is 40dB. Using this scale (Table 10-1), the sounds caused by normal breathing are found to be about 10dB, normal conversation occurs at a sound pressure of 60dB, and sounds become uncomfortable at pressures of about 120dB or 1,000,000 times threshold. This is the sound level generally heard at rock concerts.

Sound waves can travel through media other than air, but the transmission of the sound from one medium to another is usually very inefficient. When sound waves propagate from air to water, their energy is reduced by about 1000 times. This is why it is almost impossible to hear sounds initiated in air while under water. Since the transduction of sound to a neural signal occurs in the fluid medium of the inner ear, there needs to be a mechanism to enhance the transfer of sound from air to water. This is the function of the middle ear.

THE MIDDLE EAR

In order for sound to reach the auditory receptors, it must travel through the outer and middle ear to the inner ear, where the auditory receptors are found. The outer ear, consisting of the pinna and external auditory meatus, terminates at the tympanic membrane or eardrum (Fig. 10-1).

The middle ear is an air-filled cavity containing a series of three bones, called the *ossicular chain*, that transmit the sound from the eardrum to the inner ear (Fig. 10-1). For normal hearing to occur, the air pressure within the middle ear must be equal to the atmospheric pressure. These pressures are normally equalized by the eustachian tube, which connects the middle ear to the

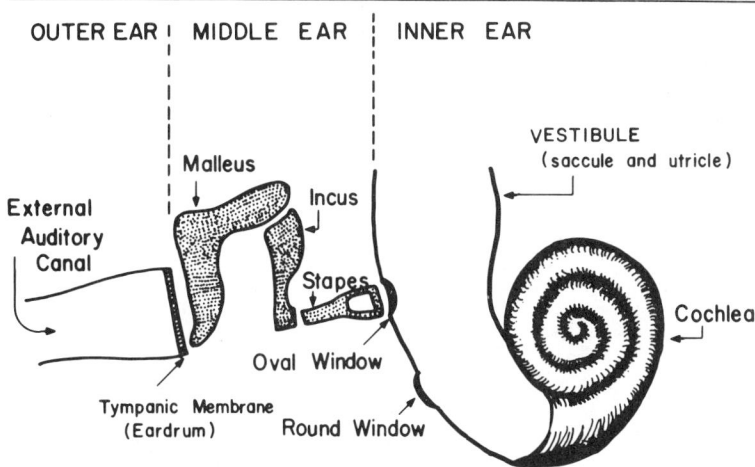

Fig. 10-1. Components of the outer, middle, and inner ear that are important in transmitting a sound wave to the auditory receptors.

pharynx and is opened when swallowing, chewing, or yawning. This is the reason that individuals chew gum or swallow repetitively when flying in an airplane, since by opening the eustachian tube the pressure within the middle ear can be reduced to the pressure of the cabin of the airplane.

The Middle Ear Bones

The three bones, or ossicles, of the middle ear are the malleus, incus, and stapes (Fig. 10-1). The manubrium of the malleus is attached to the tympanic membrane and can normally be seen by a physician during an otoscopic examination of the ear. When sound traveling through the auditory meatus reaches the eardrum, the eardrum is vibrated. This vibration is transmitted to the malleus, which in turn sets the incus and then the stapes into vibration. The footplate of the stapes sits on the oval window of the inner ear and the vibration of the stapes causes the fluid of the inner ear to vibrate.

Sound Amplification by the Middle Ear

When sound traveling in air reaches a fluid medium, very little of its energy is transmitted into the fluid; most of it is referred to by audiologists as an *impedance mismatch*. The purpose of the middle ear is to prevent this mismatch from limiting the amount of sound energy that reaches the inner ear receptors.

The middle ear prevents the loss of sound energy by amplifying sound pressures. Most of this amplification is due to the difference in the size of the eardrum and stapes. The human eardrum has an area of approximately 55 mm^2, which is about 17 times as great as the area of the stapes (3.2 mm^2). This difference causes the sound pressure to be increased by about 17 times (about 24.5dB). In addition, there is a small increase in pressure of about 1.3 times, due to the mechanical advantage gained by the leverage of the middle ear bones. Together they provide a total increase in pressure of 22 times (1.3 × 17) or 27dB. This makes up for almost all of the energy lost (30dB) due to the reflection of sound as it passes from air to water.

Not all of the frequencies in an auditory stimulus are amplified by the same amount. Figure 10-2 illustrates the gain in pressure produced by the middle ear as a function of frequency. Note that the greatest gain is achieved for sounds in the frequency range of 1,000 to 5,000 Hz, and that there are much smaller gains for frequencies above and below this level. This difference in pressure amplification at different frequencies account for our inability to hear all frequencies equally well. This is discussed in more detail later, but first we will consider how a sound stimulus that is transmitted through the middle ear is transformed into a neuronal signal by the inner ear.

THE INNER EAR

The inner ear is a series of interconnecting cavities within the temporal bone of the skull that includes the auditory cochlea as well as the three semicircular canals, the utricle, and the saccule. These latter structures are part of the vestibular apparatus and are discussed in the chapters dealing with the control of movement. The cochlea, as illustrated in Figure 10-1, is a nail-like structure containing two-and-two-thirds turns. When uncoiled, it is about 3.5 cm long.

The Cochlea

Figure 10-3 is a cross-sectional view of one cochlear coil. The outer boundary of the inner ear is called the *bony labyrinth*. Suspended within the bony labyrinth is a membrane-bound cavity called the *membranous labyrinth*. The space between the bony and membranous labyrinths is filled with a fluid called *perilymph*. It serves to cushion the membranous labyrinth much like the cerebrospinal fluid protects the brain. Like extracellular fluid, the perilymph is high in sodium and low in potassium. The membranous labyrinth, on the other hand, is filled with a fluid called the *endolymph* that is more like intracellular fluid in that it is high in potassium and low in sodium.

The cochlea is divided into three chambers by the membranes bounding the membranous labyrinth. The membranous labyrinth itself is called the *cochlear duct* or *scala media*. The upper boundary of the cochlear duct is formed by Reissner's membrane, above which is the *scala vestibuli*. The bottom of the cochlear duct is formed by

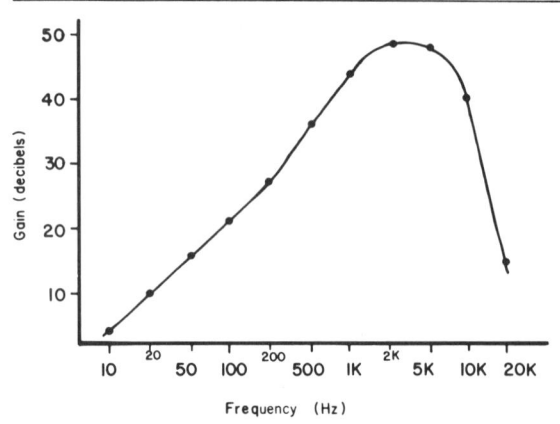

Fig. 10-2. Relationship between the frequency of sound and the amplification by the middle ear. Sounds between 1000 and 5000 Hz are amplified most. These are the frequencies of the sounds used in speech.

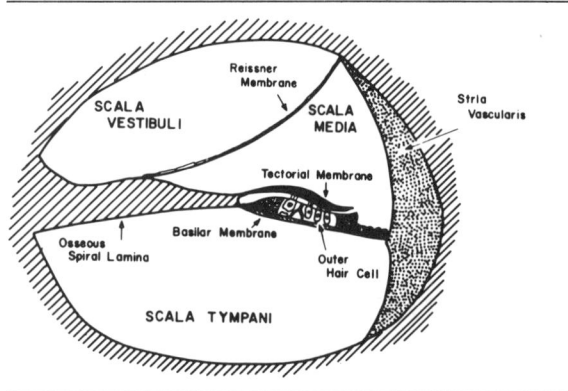

Fig. 10-3. A cross-section of the cochlea.

the *basilar membrane*, which is the boundary between the cochlear duct and the *scala tympani*. The scala vestibuli and scala tympani are interconnected with each other at the apical end of the cochlea by a hole called the *helicotrema*.

Figure 10-4A represents various aspects of the cochlea that are best appreciated by showing the cochlea as if it were uncoiled. The cochlea is widest at its base, where it borders on the middle

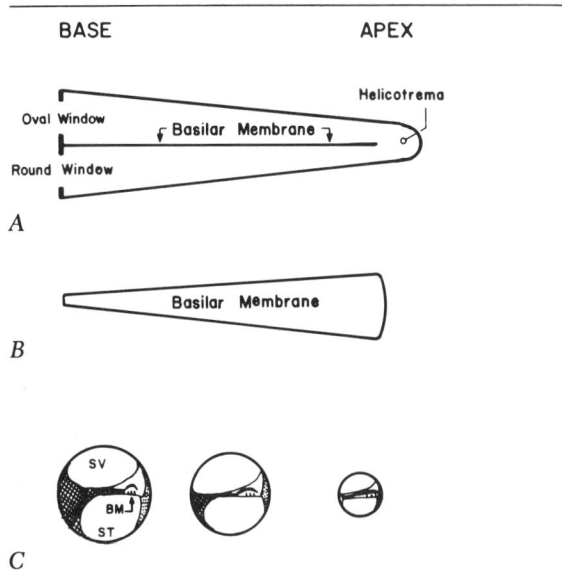

Fig. 10-4. *The difference between the basal and apical ends of the cochlea. A. The uncoiled cochlea, showing that its diameter decreases from base to apex. B. The increase in width of the basilar membrane from base to apex. C. Although the bony labyrinth decreases in diameter from base to apex, the membranous labyrinth actually becomes larger.*

ear, and is narrowest at its apex. However, the basilar membrane that forms the base of the cochlear duct is, in contrast, widest at the apex and narrowest at the base (Fig. 10-4B). The cross-sectional views of the cochlea at different positions (Fig. 10-4C) show that, although the bony labyrinth narrows from base to apex, the diameter of the membranous labyrinth actually increases from base to apex. The functional consequences of the change in basilar membrane size are discussed later, when considering the encoding of the sound stimulus by the auditory receptors.

The Cochlear Duct

From a sensory point of view, the cochlear duct is the most important chamber, because it contains the hair cells that are responsible for the transduction of a sound stimulus into a change in membrane potential (see Fig. 10-3). As indicated earlier, its upper border is formed by the Reissner's membrane. It is attached on the outer portion of the cochlea to a highly vascularized patch of tissue called the *stria vascularis*, which is responsible for maintaining the high concentration of potassium in the endolymph of the cochlear duct. The other side of Reissner's membrane is attached to a projection from the inner border of the bony labyrinth called the *osseous spiral lamina*. The lower boundary of the cochlear duct is formed by the basilar membrane, which spans the cochlea from its outer wall to the osseous spiral lamina. Resting on the basilar membrane is the *organ of Corti*, which contains the hair cells responsible for auditory transduction.

The Organ of Corti

The organ of Corti is a complex structure containing two types of sensory receptor cells, the inner and outer hair cells (Fig. 10-5). There is a single row of inner hair cells that, as their name suggests, are found on the inner surface of the cochlea near the osseous spiral lamina. In humans, there are approximately 3,400 inner hair cells along the basilar membrane from one end of the cochlea to the other. Outer hair cells are much more numerous (approximately 13,400) than inner hair cells and are organized into three rows along the basilar membrane. When viewed from above, the hairs on the inner hair cells form a single column, while those on the outer hair cells are arranged in the shape of a W.

Covering the hair cells, and forming the top of the organ of Corti, is the *tectorial membrane* (Fig. 10-5). This membrane is anchored on its inner surface to the *spiral limbus*, a mass of connective tissue lying on the osseous spiral lamina, and forms a loose connection with a group of supporting cells, called *Hensen's cells*, on the outer portion of the basilar membrane. Although the stereocilia (hairs) of the tallest outer hair cells are embedded within the tectorial membrane, it is not known if the stereocilia of the other outer hair cells or the inner hair cells are actually em-

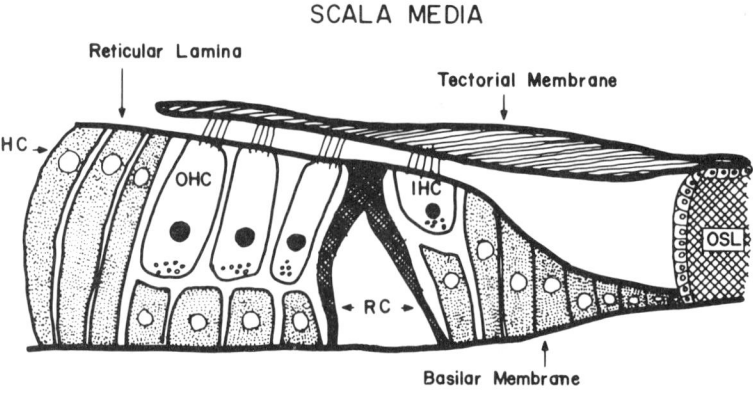

Fig. 10-5. The organ of Corti.

bedded within it. However (as discussed below), movement of the hair cells relative to the tectorial membrane causes the hairs to be bent or deformed in some way and the deformation of these hairs is responsible for the production of a receptor potential in the auditory hair cells.

The upper surface of the hair cells and the supporting cells within the organ of Corti (Fig. 10-5) join together to form a layer, the *reticular lamina*, that separates the fluid within the organ of Corti from the endolymph of the cochlear duct. Although the composition of the fluid within the organ of Corti, called the *cortilymph*, is not precisely known, it is assumed to be similar to that of the perilymph (e.g., high in sodium rather than potassium, like the endolymph).

Vibration of the Basilar Membrane

The initial step in the transduction of a sound stimulus into a receptor potential is the vibration of the basilar membrane. In order for this to occur, the sound wave must be propagated through the middle ear bones to the inner ear. When a sound stimulus causes the tympanic membrane to vibrate, the malleus, incus, and stapes also vibrate. Since the footplate of the stapes is inserted into the oval window, and since the pressure wave is amplified by the middle ear, the perilymph of the scala vestibuli is also made to vibrate.

The vibration of the perilymph causes the basilar membrane to move up and down. For low frequencies of sound, the movement of the perilymph within the scala vestibuli is transmitted through the helicotrema into the scala tympani. For higher frequencies of sound, the pressure wave is not actually transmitted through the helicotrema. Nonetheless, the vibration of the perilymph still causes the basilar membrane to move up and down. When the stapes moves in toward the helicotrema, it increases the pressure in the scala vestibuli relative to the scala tympani, causing the basilar membrane to move downward. When the stapes moves out, the pressure in the scala vestibuli is less than that in the scala tympani and the basilar membrane moves up toward the tectorial membrane.

In either case, when the vibration of the stapes pushes the oval window in, the round window must move out; conversely, when the oval window is pulled out, the round window is pulled in. Thus, in order for the basilar membrane to vibrate, there must be a pressure difference between the oval and round windows. If the pressures outside the oval and round windows were the same, the pressure wave could not be transmitted through the cochlea. Thus, the middle ear bones do more than amplify the sound pressure wave; they also transmit the sound preferentially to the oval window. When the middle ear bones do not function properly, the difference between the sound pressures reaching the oval and round windows is not nearly as great. Consequently, the loss of hearing associated with middle ear disease is greater than would be anticipated from the loss of amplification alone.

The reason why low frequency sounds pass through the helicotrema and high frequency sounds do not is related to the mechanical properties of the basilar membrane. As earlier noted, the cochlear duct and basilar membrane are narrower at the base than they are at the apex. The variation in width along the basilar membrane causes a variation in stiffness from base to apex that affects how the basilar membrane responds to different frequencies of sound. High frequency sounds cause a greater displacement of the stiffer, basal portions of the basilar membrane, whereas low frequency sounds cause a greater displacement at the more compliant, apical portions of the basilar membrane. Only when the frequency of sound is very low (less than 100 Hz) does very much energy reach the apical end of the basilar membrane and actually pass through the helicotrema to the scala tympani. The ability of a particular sound frequency to displace one area of the basilar membrane more than another is also important in the encoding of a sound stimulus by the auditory system.

THE HAIR CELLS

The sensory cells of the organ of Corti are the hair cells. These cells are excited when their hairs are bent or deformed (Fig. 10-6). The mechanism by which the movement of the hairs is converted into a receptor potential is not known. One possibility is that there are gated ionic channels within the hair cell membrane that are linked to the microfilaments inside the hairs. When the hairs are stressed, they might deform the gating molecule, causing the ionic channels to open. Another possibility is that movement of the hairs causes the release of some internal or second messenger into the cytoplasm of the hair cell, and that this molecule diffuses to the cell membrane, where it causes the opening of ionic channels.

The Receptor Potential

Although the nature of the ionic channel is not known, there is some experimental evidence supporting the view that potassium ions from the endolymph (which has a high potassium concentration) move through channels on the apical surface of the hair cell, causing depolarization. Depolarization usually is brought about by influx of sodium ions but if the electrochemical gradient for potassium is directed inward (as appears to be the case in the hair cell), an increase in potassium conductance would cause the cell to depolarize.

The advantage of producing a receptor potential in this manner is that the ions entering the hair cell do not have to be pumped out. The potassium entering the cell when it is stimulated can passively diffuse out from its basal surface into the cortilymph. At the same time, the high con-

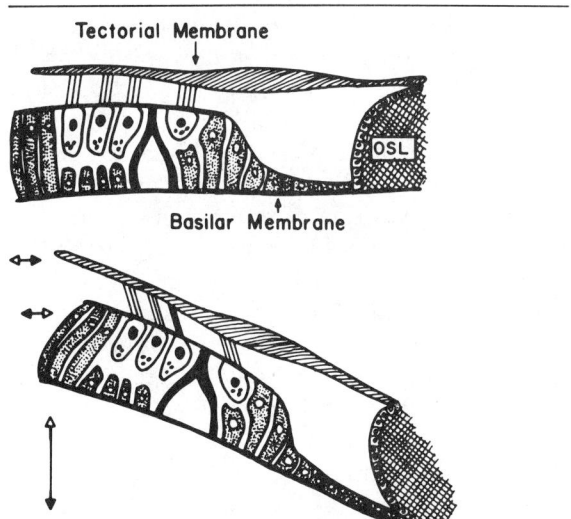

Fig. 10-6. How the hairs are bent when the basilar membrane vibrates.

centration of endolymphatic potassium needed for the production of the receptor potential is maintained by the stria vascularis. It is thought that ototoxic drugs (such as the antibiotic streptomycin) act by preventing the movement of ions through these channels. However, how they do so or why blocking the channels causes destruction of the hair cells is not known.

The Transduction Process

There is also some uncertainty about how the vibration of the basilar membrane causes the bending of the hair cells. The most likely explanation is that when the pressure difference across the cochlear duct causes the basilar membrane to move, it also causes the tectorial membrane to move (Fig. 10-6). However, because these two membranes are anchored at different sites on the osseous spiral lamina, they have a lateral movement with respect to each other so that as they are moving up and down, they are also sliding back and forth. As Figure 10-6 shows, when the basilar membrane is moving up, the tectorial membrane is pushed towards the outer hair cells, causing the hairs to be bent outward and the hair cell to depolarize. As the basilar membrane moves down, toward the scala tympani, the hairs are bent in the opposite direction and the hair cell repolarizes. As indicated previously, the hairs may not be firmly embedded in the tectorial membrane. However, even if this turns out to be the case, the hairs will still be bent because the sliding motion of the basilar and tectorial membranes causes the cortilymph to flow back and forth, bending the hairs.

The Cochlear Microphonic Potential

The receptor potential produced by all of the hair cells activated by a particular sound stimulus can be recorded with an electrode placed within the external auditory meatus, near the round window, or inside the cochlea. This potential is called the *cochlear microphonic potential* because, like the potential produced by a microphone, it is an electrical signal with a magnitude

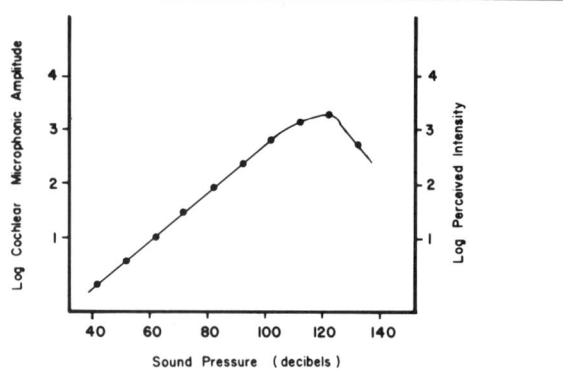

Fig. 10-7. *The relationship between the logarithm of the sound intensity and the logarithm of the electrical response is linear (left ordinate). This power law (log-log) equation (right ordinate) also describes the relationship between the stimulus intensity and the magnitude of the sensation.*

and frequency analogous to the sound stimulus. Although both the inner and outer hair cells contribute to the cochlear microphonic potential, the outer hair cells, due to their larger numbers, are primarily responsible for its production. By recording the cochlear microphonic potential, experimental evidence about the nature of the transduction process occurring within the organ of Corti can be easily obtained.

Figure 10-7 is a graph of the amplitude of the cochlear microphonic potential as a function of the stimulus magnitude. Up to a sound pressure level of approximately 120dB (the level at which sounds become uncomfortable), there is a linear relationship between the logarithm of the stimulus and the logarithm of the response. Arithmetically, this relationship can be described by the power function relating perception of sound intensity to the magnitude of the stimulus, which was discussed in Chapter 6. At stimulus magnitudes above 120dB, there is a decrease in the size of the cochlear microphonic potential, probably because the hair cells are being adversely affected by the stimulus.

Another example of the information obtained

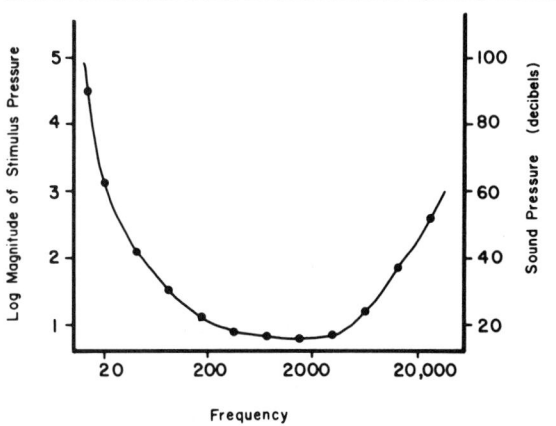

Fig. 10-8. Sounds in the frequency range between 1000 and 5000 Hz are most effective in activating the hair cells.

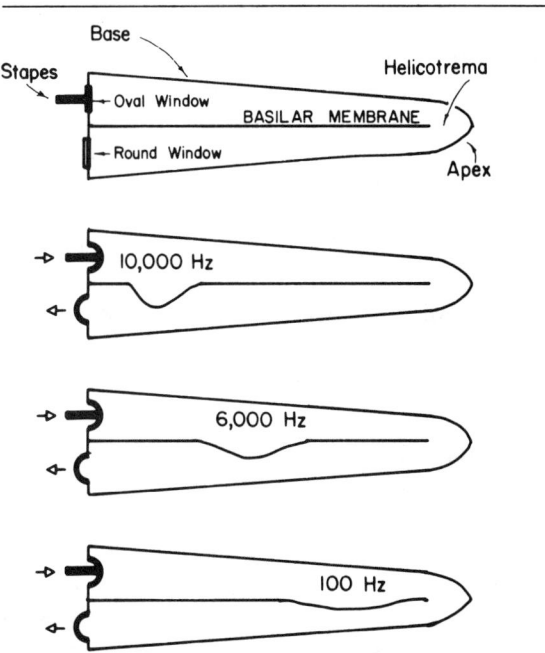

Fig. 10-9. Different frequencies of sound cause maximum vibrations at different points along the basilar membrane.

from recordings of the cochlear microphonic potential is illustrated in Figure 10-8. This graph illustrates the relationship between the magnitude of a stimulus required to produce a cochlear microphonic potential of a given size and the frequency of the stimulus. Note that the size of the stimulus is lowest for frequencies between 1000 and 10,000 Hz. This is the same range of frequencies for which the amplification of the middle ear is greatest (see Fig. 10-2) and for which the threshold for hearing is lowest. These results indicate that our ability to detect some frequencies of sound at a lower intensity than others is due to the response characteristics of the middle ear.

NEURAL ENCODING

Our discussion up to this point has dealt with the transformation of a sound stimulus into a receptor potential. Next, how the information contained in the sound is encoded by the nervous system is considered.

The hair cells of the organ of Corti are innervated by axons of the eighth cranial nerve. There are about 30,000 acoustic nerve fibers and approximately 95% of these innervate the inner hair cells. Since there are only about 3,400 inner hair cells, each cell receives an input from several sensory fibers. By contrast, there are far fewer fibers innervating the more numerous outer hair cells, so that each axon must carry information from more than one hair cell. Because of their denser innervation, it is assumed that the inner hair cells are responsible for conveying most of the auditory information from the cochlea to the central nervous system.

The Place Code for Frequency

Recall that different frequencies of sound displace different regions of the basilar membrane (Fig. 10-9); higher frequencies displace the basal portion of the basilar membrane, while lower frequencies displace the apical portions. As a result, the particular hair cell that is excited depends on where along the basilar membrane it is located. When the hair cell is depolarized, it releases a synaptic transmitter substance that excites the auditory nerve fibers innervating that cell. Thus, the nerve fiber activated by a particular frequency of sound is also dependent on where along the basilar membrane it makes synaptic contact with a hair cell. This is called the place code for frequency encoding and represents a labeled line mechanism of sensory coding.

Although it has generally been thought that the hair cells are all alike and that neural coding is entirely a function of place, recent evidence has been obtained to suggest that hair cells are also tuned both mechanically and electrically to particular frequencies. For example, hair cells located on the basilar membrane closest to the base have stiffer stereocilia than those located at the apex. In addition, they appear to depolarize more when vibrated at a high frequency than when vibrated at a low frequency. These observations would explain why our ability to detect small differences in frequencies is much greater than would be predicted from observing the vibrations of the basilar membrane, since, as illustrated in Figure 10-9, a fairly large area of the basilar membrane is displaced by a single frequency of sound.

The Auditory Pathway

The pathway by which auditory information is passed on to the cerebral cortex is fairly complex. Figure 10-10 is a simplified diagram illustrating the major components of the auditory pathway. The auditory nerve projects to neurons within the cochlear nucleus. From the cochlear nucleus, axons travel through the lateral lemniscus to the contralateral inferior colliculus. Fibers from the inferior colliculus then project to the medial geniculate nucleus of the thalamus and from there to the auditory areas (called Brodmann's areas 41 and 42) on the temporal lobe of the cerebral cortex.

Note that there are also projections from the cochlear nucleus to the superior olive in the midbrain, from which fibers join the lateral lemniscus to terminate within the inferior colliculus. In addition to the major contralateral pathways, there are also a substantial number of fibers remaining ipsilateral. The major receiving areas of the auditory pathway thus receive input from both ears. This information is used by the central nervous system to localize the direction from which a sound is coming.

The labeled-line encoding of frequency information is maintained throughout the auditory system. Within the periphery, axons originating at the apex of the cochlea (which carry information about low frequencies) are found in the center of the auditory nerve, while those from the base (which carry information about high frequency sounds) are found on the circumference of the nerve. Similarly, there is an orderly distribution of neurons within each of the relay stations, with those receiving information from the base of the cochlea separated from those receiving information from the apex. Neurons in the dorsal portion of the cochlear nucleus receive information about high-frequency sounds, while those located ventrally receive information about low-frequency sounds. Similarly, in the auditory areas of the cortex, low-frequency sounds are represented anterolaterally, while neurons that respond to high frequency sounds are localized posteromedially. This tonotopic organization of auditory information is analogous to the somatotopic organization of cutaneous sensation that was described earlier.

The Volley Code for Frequency

Information about the frequency of sounds stimulating the ear is not only encoded by a labeled-line mechanism; a frequency code is also utilized. When sounds cause the basilar membrane to vibrate, a receptor potential is produced each time the hair cells are moved up toward the tectorial membrane. If it is of sufficient magnitude, each receptor potential causes the generation of an action potential in an auditory nerve fiber. Thus, the frequency of firing in the auditory nerve fiber becomes the same as the sound used to stimulate it.

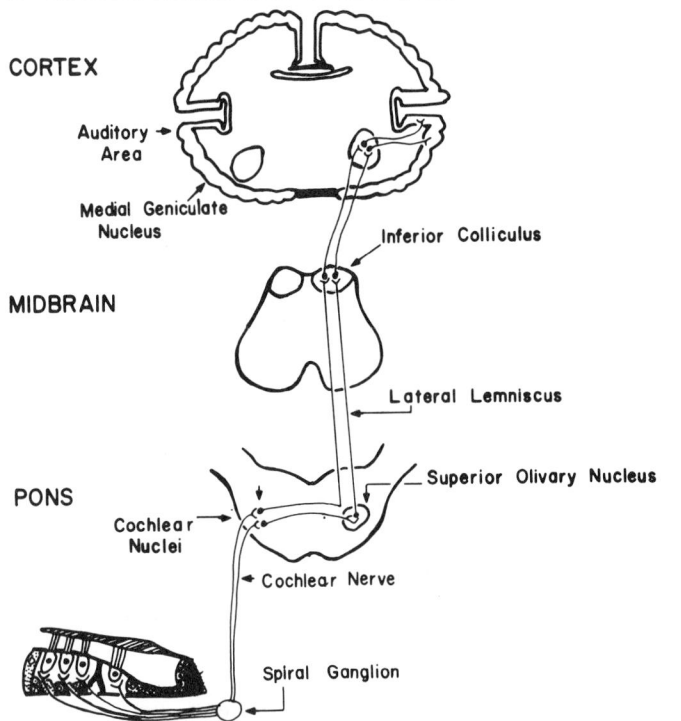

Fig. 10-10. The auditory pathway.

This is referred to as a *volley code for frequency*.

This coding mechanism cannot function in a simple way for sounds above 1000 Hz because, as discussed in Chapters 2 and 3 on action potentials, the refractory period of a neuron prevents its firing rate from exceeding approximately 1000 Hz. However, frequency encoding can be used to encode frequencies higher than 1000 Hz if more than one neuron participates in the encoding process. Even though each neuron does not fire every time the basilar membrane vibrates, at least one of them does. Since the neurons are firing in synchrony with each other, the total output from the bundle of neurons discharges at the same frequency as the sound used to stimulate the ear. This method of encoding auditory information has been shown to operate up to frequencies of approximately 5000 Hz. Above this frequency, labeled-line encoding is probably used exclusively.

Intensity Coding

If the frequency of neuronal firing is used to encode the frequency of a sound stimulus, then what mechanism is left for the encoding of auditory intensity? One way in which such encoding can be accomplished is for the nervous system to be able to interpret the number of spikes generated for each receptor potential. As the intensity of the stimulus increases, the receptor potential increases, and correspondingly the neuron fires in bursts at the peak of each receptor potential. The number of neurons which fire in response to a given frequency of sound also increases as the intensity of the stimulus is raised. This information can also be used by the central nervous system to interpret the intensity of a stimulus.

Coding for Sound Location

One of the most exquisite abilities of the auditory system is its ability to localize sounds in space, which is accomplished by feature detectors in the brain stem that receive input from both ears. Since the sound stimuli reaching the two ears are different, this information can be used by these neurons to localize the sound source. The stimuli are different in two respects. First, the intensity of the sound reaching the ear closer to the source of the sound is greater than that reaching the farther ear, due to the loss of sound energy as it passes around the head. Second, the sound wave does not reach the two ears simultaneously, because it takes longer for the sound to travel to the ear that is farther away. Both of these differences can be sensed by the central nervous system and used to determine the direction from which the sound originated.

Within the superior olive there are neurons that receive information from both ears. Those neurons that respond to high frequency sounds are sensitive to differences in the firing rate of the fibers projecting to them from the two ears. That is, they fire only in response to a particular difference in firing rate. Since sounds coming from one side of the head produce the greatest difference in intensity between the two sides, neurons within the superior olive that respond to large differences in the firing rate are feature detectors that encode sounds originating on one side of the body. Other feature detectors, responsive to different intensity differences, similarly encode sounds coming from other directions.

Other neurons within the superior olive are capable of responding to the differences between the time the sound reaches the two ears. A sound coming from one side of a head 15 cm wide reaches the farther ear about 450 microseconds after reaching the first ear. Thus, a neuron within the superior olive that only responds when action potentials propagated to it from one ear arrive 450 microseconds after the action potentials propagated from the other ear is a feature detector for sounds originating on one side of the head. Other neurons that respond to different delays are used to encode sounds originating from other locations.

Neurons responsive to differences in intensity are primarily sensitive to high frequency sounds, while those responsive to differences in the time of activation are mostly sensitive to low-frequency sounds. As it turns out, this is functionally important because the loss of energy produced by the head is much greater for high-frequency than for low-frequency sounds. Similarly, if the time for the sound to travel from one ear to the other exceeds the period ($1/f$) of the sound wave, then the time difference is not resolvable. For example, if the sound originates on one side of the head and the time delay is 450 microseconds, then the frequency of sound cannot exceed 2,200 Hz. The information encoded by these neurons in the superior olive is projected along the auditory pathway to the auditory cortex, where the sensation of sound localization is produced.

CLINICAL ASPECTS OF HEARING
Types of Hearing Loss

Individuals who suffer from hearing loss due to physical damage of the auditory apparatus are divided into two categories: those with conductive loss and those with sensori-neural impairment. Conductive hearing loss results from an inability to conduct the sound stimulus into the inner ear. This can result from the build up of ear wax (cerumen) in the outer ear, preventing the sound from reaching the tympanic membrane, or from otitis media, an infection of the middle ear, in which a build-up of fluid around the ossicular chain prevents the transmission of the sound from the tympanic membrane to the oval window. The treatment of conductive hearing loss is often very simple. For example, removing the cerumen plug in the ear canal or draining the fluid from the middle ear cavity can produce rapid and dramatic improvement in hearing. Moreover, even if the hearing loss cannot be entirely corrected, simple amplification of the sound will make it possible for the individual to hear normally.

By contrast, an individual suffering from sensori-neural loss cannot gain much improvement from a hearing aid that simply amplifies sound because the sensori-neural deficit results from damage to the inner ear where sensory encoding takes place. For example, as an individual ages his or her ability to hear high frequency sounds diminishes. This condition is called *presbycusis* and is due to the gradual loss of hair cells at the basal end of the cochlea. Because part of the auditory spectrum is not heard well, an inability to distinguish sounds from each other develops. This is particulary evident when holding a conversation with an elderly individual who, because of his or her hearing loss, continually asks for phrases to be repeated. An inability to encode the high frequency sounds made by consonants causes the lack of understanding. Speaking louder does not help. In fact, it may make the problem worse because individuals with sensori-neural deficits often perceive sounds with a greater than normal loudness. In addition to the loss of ability to encode high frequency sounds, these patients also report the presence of a continuous ringing or buzzing sound in their ears. This perception of sound in the absence of a sensory signal is called *tinnitus*. Generally the frequency of the tinnitus is in the range of 2 to 8 kHz (i.e., in the same range as that of sounds for which there is a deficit in sensory encoding).

Sensori-neural deficits may result from trauma, diseases, and reactions to drugs such as quinine or streptomycin. The most common cause of hearing loss is exposure to environmental noise. Individuals who work in factories, airports, and other loud places have a higher than normal incidence of sensori-neural hearing impairments. Interestingly, all of us may suffer the accumulative effects of a noisy society because it has been shown that the hearing loss associated with presbycusis of urban dwellers is much greater than that of rural populations.

Tests for Hearing Loss
Clinically, conductive hearing loss can be distinguished from sensori-neural loss on the basis of simple tuning fork test called the Rinne test. The test is performed by placing a vibrating tuning fork near the ear and asking the patient to indicate when the sound disappears. The base of the still vibrating tuning fork is then placed against the patient's mastoid bone. If the hearing loss is due to a conduction deficit, the sound will again be heard because it will be conducted through the bones of the skull to the intact inner ear. If the loss is due to a problem with the inner ear, however, the air-conducted sounds will be heard much longer than the bone-conducted sounds and the patient will not report hearing the tuning fork when it is placed against his or her mastoid bone.

The Rinne test is not very conclusive unless the hearing loss is fairly severe because air conduction is much more sensitive than bone conduction. Thus a patient with a mild conductive loss might still hear the tuning fork longer when held near the ear than when placed on the mastoid. Another ambiguity in interpreting the Rinne test occurs when there is a unilateral deficit. For example, a patient with a unilateral sensori-neural hearing loss may report hearing the bone-conducted vibrations longer than those produced in air because the sounds are being conducted through the skull to the undamaged ear.

The results of the Rinne test can be confirmed with another simple tuning fork test called the Weber test. In this test, the base of the vibrating tuning fork is placed on the midline of the skull and the patient is asked to indicate the ear in which the sound is heard best. If there is a unilateral conductive lesion, the sound will be heard better in the ear with the deficit. This paradoxical result occurs because the sound is conducted through the bones to both ears equally and thus appears louder in the ear that normally does not hear as well. In the presence of a sensori-neural hearing loss, the sound is heard better in the ear without the deficit.

Quantitative information about the extent of a hearing loss can be obtained by audiometric techniques in which pure tones of known frequency are presented to the patient and his or her hearing threshold is determined. Figure 10-8 is typical of the results that are obtained during an audiometry test of a normal individual. If there is a hearing loss, its magnitude can be judged by the increase in stimulus intensity (measured in decibels) required for the sound to be detected at each of the frequencies in the normal hearing range.

III Motor Control

11 Overview of the Motor Control System

The components of the motor control system are diagrammed in Figure 11-1. Notice in particular the extensive interconnections among the components. These feedback loops are used extensively by the motor control system, in both the initiation and coordination of muscle activity.

Even the simplest of voluntary movements requires the participation of all these control centers for its proper performance. Very briefly, the "idea" for a movement is generated in the association areas of the cerebral cortex and then is transmitted to the motor cortex for execution. To carry out the movement, a "command" is sent from the motor cortex to the spinal cord via the extrapyramidal and pyramidal motor systems. The cerebellum receives information from the cortex about the "intent" of the movement prior to its execution, and from the spinal cord about how well the movement is being performed. The cerebellum uses this information to aid in the generation of the initial motor command and to alter the command while the movement is in progress. The basal ganglia contain the circuitry for a variety of complex movements that can be called upon to adjust bodily position in accordance with the movement being executed. Finally, in the brain stem, the vestibular apparatus and its nuclei maintain equilibrium and balance, while the reticular formation maintains tone in the antigravity muscle so that the body can maintain an upright posture.

This chapter provides an overview of how each of the components of the motor control system mentioned above work together to produce coordinated movements. In subsequent chapters, these components are described in more detail.

All muscular activity, whether initiated by the central nervous system or as a response to a peripheral stimulus, is generated by the firing of the alpha motoneurons that innervate skeletal muscle. These alpha motoneurons are referred to as the *final common pathway* for motor activity, because all of the components of the motor control system ultimately converge on them and act through them to produce their effects on movement.

THE MOTONEURON AND MOTOR UNIT

The alpha motoneurons that innervate a particular muscle are grouped together within the ventral horn of the spinal cord or cranial nuclei to form a motor neuron pool. Each neuron within the pool branches to innervate a group of muscle fibers within the muscle. The fibers innervated by a single alpha motoneuron are called a *motor unit*. Motoneurons and motor units vary in size.

The number of muscle fibers in a motor unit is proportional to the diameter of the alpha motoneuron innervating it, that is, the larger the alpha motoneuron, the greater the number of muscle fibers it innervates. In general, when a motor neuron pool is stimulated, the smaller motoneurons are fired first, meaning that the smaller motor units are activated before the larger ones. This is called the *size principle*. The functional consequences of this sequence of motoneuron activation (from small to large) is described in the next chapter, where the properties of motor units are described.

SPINAL CORD REFLEXES

Reflexes are motor behaviors that occur automatically in response to a stimulus. The neuronal circuitry required for the performance of a reflex behavior is genetically determined and built into the nervous system during development of the fetus. However, reflexes should not be thought of as immutable, rigid behaviors that always occur in the same way. Learning can alter the neuronal connections of old reflexes and establish new ones. The neuronal centers responsible for gener-

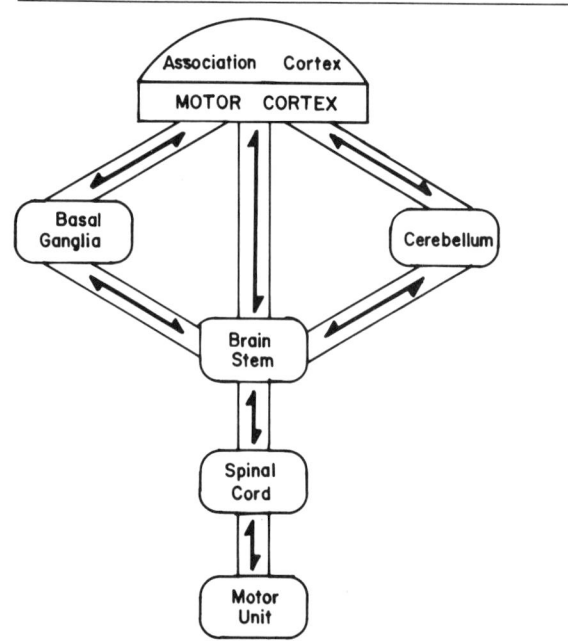

Fig. 11-1. The interconnections of the components of the motor control system.

ating the reflex are subject to control by higher centers of the central nervous system. As a result, ongoing behaviors can greatly influence the conditions under which a reflex is evoked and also the type of behavior that it produces. In its simplest form, a reflex consists of an afferent or sensory input, an efferent or motor output, and a central integrating center.

There are two major sets of sensory inputs participating in spinal cord reflexes. One set comes from the skeletal musculature and the other comes from the skin. Sensory inputs coming from the muscle fibers originate within the Golgi tendon organ and muscle spindle and detect muscle tension and length, respectively.

The muscle spindle receptors are innervated by small motoneurons, called *gamma motoneurons*. They are interspersed among the alpha motoneurons that innervate the muscle containing the muscle spindle. The gamma motoneurons play an important role in the control of spinal cord reflexes.

The other major sensory input to the spinal cord comes from the cutaneous receptors. The most important of these are pain neurons. When activated, these neurons initiate a reflex that rapidly removes the body from the painful stimulus.

The integrating center for spinal cord reflexes is the *motoneuron pool*. Most of the sensory fibers from outside the central nervous system, as well as those within the central nervous system that influence the activity of the alpha and gamma motoneurons within the motoneuron pool, do not do so directly. Instead, they synapse with interneurons that make connections with the motoneurons through a variety of pathways.

The complex network of neurons within the integrating center of spinal cord reflexes allows the central nervous system a great deal of flexibility in directing the behavior of alpha motoneurons. One of the most important requirements for the smooth performance of a movement is the inhibition of the muscles that act to antagonize the movement. That is, if the arm is to be extended, the flexor muscles must be inhibited. This inhibition is performed automatically by the neuronal circuitry within the spinal cord. The same axon that generates activity in one motoneuron pool, whether from the peripheral or from the central nervous system, also sends a collateral to an interneuron pool that in turn sends inhibitory fibers to the motor neuron pool innervating the antagonistic muscle. In the absence of this interneuronal circuitry, two central axons, one going to the agonist and one going to the antagonist, would be required for every movement, doubling the number of neurons required to coordinate a movement.

THE BRAIN STEM

Another input to the motor neuron pool, indicated in Figure 11-1, is the brain stem. This is the major processing center through which the motor control system sends information to the spinal cord. The brain stem receives input from the cerebral cortex, the basal ganglia, the cerebellum, and the vestibular apparatus. It also receives an extensive sensory input, making it an ideal center for the integration of sensory and motor information. The brain stem motor area is part of the reticular formation. It is divided into excitatory and inhibitory areas. The spasticity that results from a cerebral stroke is a consequence of losing cortical control over the brain stem reticular excitatory area.

In addition to its input to the reticular formation, the vestibular apparatus has a direct pathway to the motor neuron pools, and participates in a number of reflexes that permit balance to be maintained during movement. Although it is not shown in Figure 11-1, the vestibular system is also responsible for the coordination of head and eye movements, so that visual fixation on an object is not lost when the head moves.

THE BASAL GANGLIA AND CEREBELLUM

Figure 11-1 also shows two important components of the motor control system that do not have direct input to the spinal cord. These are the basal ganglia and the cerebellum. The basal ganglia are a group of subcortical nuclei whose function is not well understood but that appear to be involved in the production of automatic postural adjustments that accompany movement. Lesions within the basal ganglia produce characteristic and often debilitating motor disturbances.

The primary role of the cerebellum is to coordinate the sequence and timing of motoneuron firing, so that smooth movements can be produced. The cerebellum also plays a role in the regulation of posture, along with the vestibular system. Although a tremendous amount of information is available about the neuronal circuitry of the cerebellum and how it is influenced by the extensive sensory input it receives, there is as yet no real understanding of how it performs its function.

THE CEREBRAL CORTEX

The final input to the motor neuron pool comes from the cerebral cortex. It is the highest level of the motor control system and is responsible for movements requiring the greatest degree of dexterity. There are two major systems used by the cortex to control movement. One of these is the pyramidal system, which includes the corticospinal and corticobulbar tracts. Although the corticobulbar tract does not pass through the pyramids as the corticospinal tract does, it has the same relationship to the neurons of the cranial nerve nuclei as the corticospinal tract does to the alpha motoneurons.

All of the other cerebral cortical neurons that influence the activity of spinal motoneurons do so via the reticular formation and are grouped together as the extrapyramidal system. Previous terminologies have grouped the basal ganglia together as the extrapyramidal system, and these two usages of the term should not be confused. The extrapyramidal system originating from the cortex serves two major functions, one of which is to regulate muscle tone so that the body is maintained in an upright posture. The other function is to provide a pathway for the cortex to direct the production of movement. It performs the same role in this regard as the pyramidal system, but is not capable of duplicating the finely tuned movements of the fingers and muscles of articulation that can be produced by the pyramidal tract system.

This brief outline of the motor control system is meant to provide an overview of its components, so that as each is described in detail it can be placed in its proper context. Chapter 12 begins with a discussion of the spinal motoneurons and the skeletal muscle fibers they innervate.

12 Alpha Motoneurons and Motor Units

The fundamental role of the motor control system is to produce purposeful movements, which is accomplished by contracting the appropriate muscle fibers in the proper sequence, with the force necessary to carry out the movement. Although a great deal of the organization required to carry out this task occurs within the components of the motor control system in the brain, a significant amount of control takes place within the spinal cord. This chapter describes how the physiological properties of the spinal alpha motoneurons, and the motor units they innervate, contribute to the production of smooth, coordinated movements that are necessary to carry out the tasks assigned to the muscles by the motor cortex.

ALPHA MOTONEURONS

Alpha motoneurons are organized into motor neuron pools or nuclei within the ventral horn of the spinal cord. Motor neuron pools innervating axial and trunk muscles lie medially to those innervating the limb muscles. Moreover, those motor neuron pools innervating extensor muscles are located ventrally to those innervating extensor muscles. This anatomical segregation of motor neuron pools innervating different functional groups of muscles simplifies the task of ensuring that afferent fibers synapse on the appropriate motoneurons. This is one of many examples in which the anatomical organization of the motor cortical system contributes significantly to the performance of motor activity. Each motoneuron of a particular motor neuron pool innervates muscle fibers that are contained within the same anatomically defined muscle. All those muscle fibers innervated by a single alpha motoneuron are defined as a motor unit. We will describe the properties of the motoneurons first, the properties of the motor units next, and conclude by showing how these properties aid in producing movements.

As explained in Chapter 4, Synaptic Transmission, action potentials, generated in the initial segment (axon hillock) of the motoneuron, are propagated down the axon. The shape of the motoneuron action potential is different from that of nerves, in that the motoneuron action potential has a large and prolonged afterhyperpolarization. This afterhyperpolarization, due to an increase in potassium conductance, serves to limit the frequency of motoneuron firing to about 30 impulses per second.

Motor neurons vary in size from 30 microns to 70 microns. The largest of these have the largest axons and the largest motor units. The cell body, or *soma*, and dendrites are covered with synaptic terminals. About 80% of the 20,000 or so synapses formed on the largest alpha motoneurons are found on the proximal portions of the dendritic trees. Axons that originate in different neuronal groups synapse on different regions of the motoneuron surface. For example, those axons that come from the Ia fibers of the muscle spindle form excitatory synapses on dendrites, while those coming from inhibitory interneurons tend to form synapses on the cell body, giving the inhibitory systems a geometric advantage in controlling the excitability of the alpha motoneuron, since they are closest to the axon hillock where action potentials are generated.

The variation in motoneuron size is important to the functional organization of the motor neuron pool, because small motoneurons are more excitable than large ones and are thus recruited first during most movements. However, before considering the role of motoneuron size in the motor control system, we will describe the various types of motor units and the alpha motoneurons they are associated with.

MOTOR UNITS

The motor unit is the smallest anatomical element that can be used by the motor control system to carry out a movement, because whenever an alpha motoneuron is fired, all the muscle fibers it innervates are activated. But the motor unit is not simply an anatomical grouping of muscle fibers; it has a functional value as well. The muscle fibers that constitute a given motor unit are scattered throughout the muscle, so that the force generated by the activation of a single

Table 12-1. *Characteristics of muscle fibers*

Property	Classification		
	FF (FG, IIb, white)	FR (FOG, IIa, red)	S (SO, I, red)
Speed of shortening	fast	fast	slow
Resistance to fatigue	poor	intermediate	high
Fiber diameter	large	intermediate	small
Size of motor unit	large	intermediate	small
Sarcoplasmic reticulum	large	intermediate	small
Vascular supply	sparse	intermediate	dense
Myoglobin content	none	little	high
Glycogen content	high	intermediate	low
Mitochondria	few	intermediate	many
Glycolytic enzymes	high	intermediate	low
Oxidative enzymes	low	intermediate	high
Myosin ATPase	high	high	low

motor unit is distributed over a large part of the muscle. All of the muscle fibers within a muscle unit share the same physiological properties, and therefore the contractile activity of the motor unit is uniform. An understanding of the characteristics that distinguish one motor unit type from another is necessary to appreciate the significance of the functional organization of the spinal cord motor neuron pools.

Classification of Muscle Fibers

Muscle fibers are divided into three groups based on their histological, biochemical, and physiological properties. Table 12-1 lists the characteristics of the different types of muscles and indicates how they are named under different classification schemes.

The most physiologically oriented of these schemes divides muscles on the basis of their contractile speed and fatigability. *Fast fatigable (FF) muscles*, as their name implies, contract quickly and fatigue easily. They derive most of their ATP from glycolytic enzymes, have few mitochondria, high myosin ATPase activity, a low capillary supply, and are the largest of the muscle fibers. *The slow (S) fibers*, by contrast, have long contractile times, obtain their ATP from oxidative enzymes, have many mitochondria, low myosin ATPase activity, a rich capillary supply, and are the smallest of the muscle fibers. The *fatigue resistant (FR) fibers* have characteristics that are in between those of the FF and S types.

The interrelationship of the qualities that distinguish these muscle types is quite obvious. The fast contractile times of the FF fibers result from the rapid cross-bridge cycling rate indicated by their high myosin ATPase activity. Because of their large size, diffusion of oxygen is limited, and they must rely on anaerobic metabolism; hence, they have a high concentration of glycolytic enzymes. Since oxygen needs are low, their capillary supply need not be extensive. Finally, because of their reliance on glycolytic metabolism, FF fibers fatigue easily when the rather limited store of glycogen within the muscle is exhausted.

Similarly, the low myosin ATPase activity of the S fibers indicates a slow cross-bridge cycling rate and long contraction time. Because these fibers are small, diffusion of oxygen is not a limiting factor. Thus, the dense capillary supply coupled with the oxidative enzymes needed to make use of the oxygen delivered allows S fibers to contract for long periods of time without fatigue.

FF fibers are also classified as FG (fast glycolytic) and as group IIb fibers, while the S fibers have also been classified as SO (slow oxidative) and as group I fibers. The intermediate fibers have been called FOG (fast oxidative-glycolytic) and group IIa fibers, in addition to the FR classification indicated previously.

Most muscles contain a mixture of all three fiber types. Those muscles, such as the gastrocnemius, that contain a preponderance of FF and FR fibers appear white. Those muscles, such as the postural muscles, that contain mostly S fibers appear red. This gives rise to yet another classification method, which divides muscles into red muscles, which are slow and nonfatigable, and white muscles, which are fast and easily fatigable. The red color of the slow muscles is due to the high resting blood flow, high capillary density, and high concentration of myoglobin within the muscle fibers. This text uses the FF, FR, S scheme because it is based on the physiological characteristics of the muscle fibers.

The different muscle fiber types are adapted to serve different needs. Clearly, the S-type fibers are ideal for muscles that must undergo long periods of contractile activity. They require less ATP for force production (because of their low myosin ATPase activity) and, as long as the oxygen supply is adequate, they are essentially nonfatigable. In exchange for this characteristic, however, their shortening velocity is slowed and their force production is limited. On the other hand, because the FF fibers are larger, they can

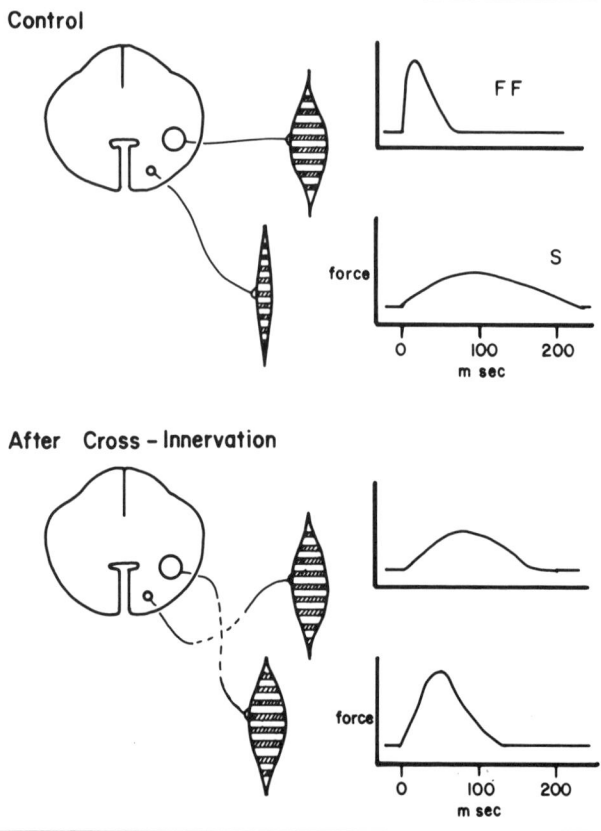

Fig. 12-1. The physiologic characteristics of a muscle fiber are determined by the pattern of activity in the alpha motoneuron innervating it. This can be demonstrated by a cross-innervation experiment, in which the alpha motoneurons innervating fast (FF) and slow (S) muscle fibers are interchanged. After the new connections are established, the contractile speed and many of the physiologic characteristics of the muscle change to correspond with its new innervation.

produce more force. However, they can only act for a short period of time before they fatigue. The challange for the motor control system is to choose the proper muscle fiber type for the required movement. How this problem is solved is explained after a description of the development of muscle fiber types.

Determination of Muscle Fiber Types

At birth, all muscle fibers appear to be of the S type. Over the first few months after birth, the muscle fiber types begin to differentiate. Since all of the muscle fibers in a particular motor unit develop the same physiological characteristics, it was assumed that its motoneuron was responsible for determining the type of muscle fiber that develops. This was tested by interchanging the motor nerves going to a fast and a slow muscle.

Figure 12-1 illustrates a cross-innervation experiment, in which the nerves going to a red (slow) and a white (fast) muscle were cut and then reattached with the nerve normally going to the fast muscle tied to the distal end of the nerve normally going to the slow muscle. The nerve that had normally innervated the slow muscle was tied to the distal end of the nerve going to the fast muscle. After regeneration of the nerves occurred, the fast muscle took on the characteristics of a slow muscle while the slow muscle took on the characteristics of a fast muscle.

This process can occur pathophysiologically. If a motor nerve is damaged by disease or trauma, the muscle fibers it innervated atrophy. In time, the nerve regenerates, but in doing so, the muscle fibers it then innervates are all close to each other rather than being spread throughout the muscle, as is normally the case. Biopsies of these muscles show that local groups of similarly typed muscle fibers are formed in the region of the regenerated nerves, demonstrating that the nerves determine the type of muscle fiber that develops. It has been shown by a variety of experiments that it is the activity of the motoneuron rather than the type of motoneuron that determines the type of muscle fiber which develops. For example, if a cross-innervation experiment is carried out so that a fast nerve is made to innervate a slow muscle, it would normally change the muscle into a fast muscle. But if the nerve is continuously stimulated electrically at a rate of about 10 stimulations/second (the normal rate of firing for nerves innervating slow muscles), the slow muscle retains its normal characteristics. In other words, it is the activity of the alpha motoneuron, rather than any trophic substance it may release, that causes a particular muscle fiber type to develop.

Muscle fiber characteristics can also be altered under physiological conditions that increase the usage of all motor units, such as exercise. During endurance training, when large groups of muscles are used for long durations, there is an increase in the ability of muscle fibers to utilize oxidative metabolism to produce ATP. Accordingly, capillary density, mitochondria, oxidative enzymes, and fatigue resistance all increase. Whether these changes represent an actual conversion from FF or FR to S fibers is not clear. However, the ability of the alpha motoneuron to determine the type of muscle fiber that develops is important to the ability of the motor control system to choose the proper motor unit to perform a particular movement.

PHYSIOLOGICAL CONTROL OF MOTOR UNITS

In addition to the variation in physiological properties, motor units vary greatly in size. Some alpha motoneurons innervate only a few muscle fibers, while others may innervate well over a thousand. The motor control system thus must not only choose the right type of motor unit, but must also choose the motor unit of the appropriate size. As it turns out, the motor control system meets this requirement in a simple and elegant way. Rather than requiring the central nervous system to pick and choose among all of the alpha motoneurons in a motor neuron pool, the right motoneuron is chosen automatically.

The Size Principle

Small motoneurons are easier to excite than large ones, because the membrane resistance of the small cells is greater than that of the large ones. Recall from the chapters on synaptic transmission that the size of the voltage change produced by the action of synaptic transmitter is proportional to the input resistance of the cells and that the input resistance increases as cell size decreases. Thus, any given excitable synaptic input produces a greater depolarization in a small cell than in a large cell. Because of this phenomenon, small alpha motoneurons are recruited before large ones, whether the command for activity comes from the motor cortex, the brain stem, or the afferent limb of a reflex. This orderly recruitment of motoneurons according to size is called the *size principle*.

Because of the size principle, the sequence of motor units activated during a movement is always the same, with the small ones (innervated by the small alpha motoneurons) recruited before the large ones. Therefore, when the need is for a fairly fine movement that requires only a small amount of force, the motor control system does not have to specifically choose a small motor unit to accomplish the task. Instead, it provides a weak input to the spinal motoneuron pool; because of the size principle the small motoneurons are automatically chosen. If the force required is increased, then the motor control system simply increases the strength of its input to the motoneuron pool and the larger motor units are activated, adding to the force of the contraction.

The size principle, along with the phenomenon by which the activity of the alpha motoneuron determines the physiological characteristics of the muscle fiber it innervates, also allows the motor control system to easily choose the proper type of motor unit for a sustained muscular effort.

Because small motoneurons are always activated first, they participate in all movements requiring their motor neuron pools. This means that the muscles they innervate are always active, and these muscles consequently develop fatigue resistant properties. Thus, if a fatigue resistant muscle is required to sustain a movement (such as holding the head erect) for a long period of time, the motor control system does not have to "know" which alpha motoneuron innervates an S type fiber. If it provides a small input to the motor neuron pool, only the small alpha motoneurons are excited. Since these are used in all movements, the muscle fibers they innervate will have developed fatigue resistance and are able to sustain the activity without fatiguing. If the contractile force is not sufficient, additional motor units have to be activated. Because of the size principle, these units are automatically selected on the basis of their ability to withstand fatigue.

The preceding discussion illustrates a general principle by which the motor control system is organized. In order to reduce the number of neurons required to produce activity, a large number of patterned responses are built into the motor control system. Because of this, the motor cortex can carry out many movements by issuing a command to subcortical neuronal circuits. These then carry out the motor behavior automatically, using the neuronal circuits built into the nervous system. The following chapters provide a number of other examples of how the neuronal circuitry of the motor control system aids in the production of movement.

13 Spinal Cord Reflexes

A reflex is a simple motor response to an environmental stimulus. For example, touching a hot stove results in the initiation of a withdrawal reflex that causes your hand to be removed from the stove before you are aware of the pain. The coupling of the stimulus to the response is based on the neuronal circuitry built into the nervous system. The circuitry consists of a receptor, an integrating center, and an effector. When the receptor is stimulated, it initiates a chain of neuronal activity that ultimately results in the activation of the effector.

However, the response is by no means invariant. It can be altered by a variety of inputs to the reflex integrating center that affect the excitability of the alpha motoneuron and thus the magnitude of the response. Reflexes are also used by the motor control system to generate and coordinate movements. The presence of circuitry built into the nervous system to produce well-defined responses makes it possible for the cerebral cortex to produce movements in a highly efficient manner. This chapter describes the receptors involved in the production of spinal cord reflexes, the circuitry that makes the reflexes possible, and some aspects of their functional significance.

MUSCLE RECEPTORS
The importance of muscle receptors (the Golgi tendon organs and the muscle spindles) in the control of muscle activity becomes apparent when the number of sensory and motor fibers innervating the muscle is compared. For example, the biceps muscle, which is innervated by about 1000 alpha motoneurons, has approximately 300 Golgi tendon organs and an equal number of muscle spindles. These receptors are innervated by well over 1000 afferent and efferent fibers. They play an important role in the control of

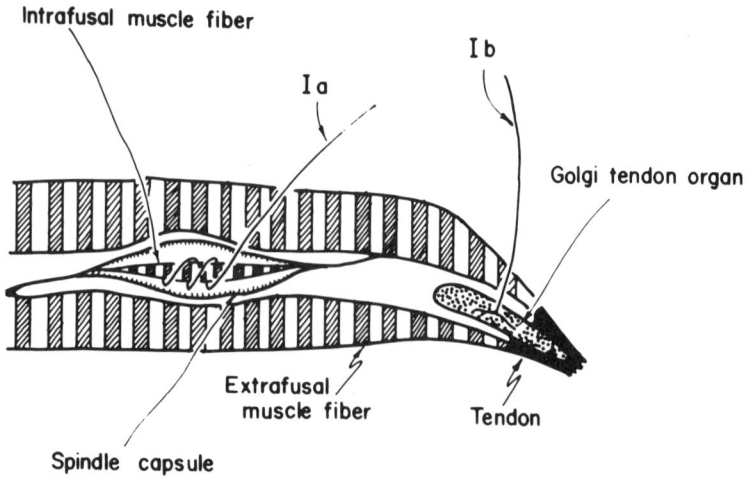

Fig. 13-1. The muscle receptors involved in the reflex control of movement.

movement and in kinesthesia (the awareness of the position of our limbs in space).

Golgi Tendon Organs

Golgi tendon organs are encapsulated sensory receptors, about 0.5 to 1.0 mm in length and 100 microns in diameter, found at the tendonous origins and insertions of a muscle (Fig. 13-1). They are innervated by large sensory afferents, called *Ib afferents*, that enter the capsule and divide to send small unmyelinated branches to tendonous fibers within the capsule. When the muscle contracts, it stretches the tendon and causes the nerve terminals to be distorted. This distortion gives rise to a generator potential that in turn initiates action potentials in the Ib afferent fibers.

Only a few muscle fibers attach to a single tendon organ, permitting each Ib afferent to convey information about the activity of a discrete group of motor units. The receptors are extremely sensitive, so that the force generated by a twitch of a single motor unit is sufficient to activate the sensory endings. As the force of contraction is increased, the frequency of firing in the Ib afferent is increased.

The anatomic relationship between the muscle and the muscle receptors is important in their function. For example, the tendon organs are arranged in series with the muscle, which means that when the muscle contracts, the tendon organ is stretched and the receptor activated. This is in contrast to the muscle spindles, which because they are arranged in parallel with the muscle, cease their activity when the muscle contracts.

The Muscle Spindle

The muscle spindle is a far more elaborate receptor than the Golgi tendon organ, in terms of both its structure and its function. Not only does the muscle spindle contain several types of sensory endings, it also is innervated by several types of afferent fibers, and most importantly, receives an efferent innervation capable of controlling its sensitivity.

The muscle spindle (Fig. 13-2) is an encapsulated receptor containing a small number (usually 5–15) of specialized muscle fibers that are innervated by both sensory and motor fibers. These muscle fibers are called *intrafusal muscle fibers* because they are within the muscle spindle. The ordinary muscle fibers that are used to generate force are, by contrast, called *extrafusal muscle fibers*.

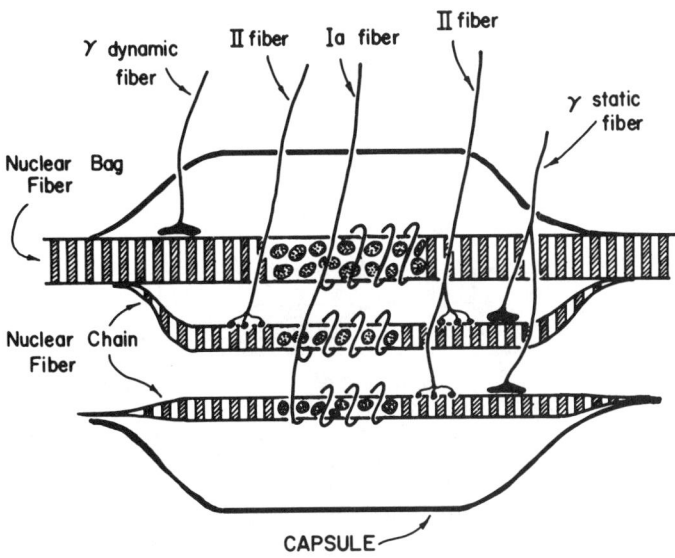

Fig. 13-2. The muscle spindle contains two types of intrafusal muscle fibers, nuclear bag and nuclear chain fibers. Gamma dynamic and gamma static motoneurons control the length of the intrafusal fibers. Sensory information is transmitted from the muscle spindle by group I and group II afferent fibers.

The Intrafusal Fibers

There are two major types of intrafusal muscle fibers, called *nuclear bag* and *nuclear chain fibers*. Their names derive from the arrangement of nuclei within them. The nuclei of the nuclear bag fibers are grouped together so that several nuclei can be seen in cross-sectional slices of the muscle fiber. In the nuclear chain fibers, the nuclei are arranged in single file so that only one nucleus can be seen in a cross-section.

There are other histological differences as well. The nuclear bag fibers are about twice as large (16 microns in diameter and 8 mm long) as the nuclear chain fibers. Usually there are 1 to 3 bag fibers and 4 to 6 chain fibers within a muscle spindle. As Figure 13-2 indicates, the nuclear chain fibers typically merge with the nuclear bag fibers and end within the capsule, while the bag fibers extend beyond where the capsule merges with the connective tissue of the muscle.

Recently, the nuclear bag fibers have been subdivided into two groups: bag_1 and bag_2 fibers. It is not necessary to go into the details of how these two fibers differ, except to point out that the bag_2 fibers appear to share many of the histological and physiological features of the chain fibers.

The Muscle Spindle Afferents

Two types of afferent fibers emerge from the muscle spindle. The largest, called the *Ia afferent* or *primary ending*, branches within the spindle to innervate each of the intrafusal muscle fibers. As indicated in Figure 13-2, the Ia sensory fiber terminates as a spiral or ring around the central region of the intrafusal muscle fiber, where the nuclei are found. Smaller afferent fibers, called *group II* or *secondary endings*, primarily innervate nuclear chain fibers. These afferent terminals are formed on either side of the primary afferent endings. Although each spindle receives only one primary afferent fiber, several secondary fibers generally enter each spindle.

The Gamma Motoneurons

The motor innervation of the muscle spindle is extremely important. About 10 branches of motor fibers, called *gamma efferents*, enter each spindle to form multiple endings on the intrafusal muscle fibers. The motor fibers can be divided into two functional groups, called the *gamma dynamic* and *gamma static fibers*. The reason for naming one of these groups dynamic and the other static will be explained shortly. For now, note that Figure 13-2 shows the gamma dynamic fiber going to the nuclear bag fiber and the gamma static fiber going to the nuclear chain fiber. Although this dichotomy is controversial, it is reasonably correct and permits us to explain a number of important functional features of the muscle spindle.

Since the muscle spindle is arranged in parallel with the extrafusal muscle fibers, when the muscle is stretched, the intrafusal fibers are also stretched, causing the primary and secondary endings to be activated. The primary ending fires vigorously during the stretch, then reduces its firing rate (or adapts) after the stretch is completed, while the secondary ending increases its firing rate in proportion to the amount of stretch. These differences in responsiveness allow the primary ending to convey information about both the velocity and magnitude of a stretch (recall the discussion of sensory coding in Chapter 6), while the secondary endings can only indicate the amount of stretch.

Unloading of Muscle Spindles

When the muscle contracts, the activity of both receptors decreases. Figure 13-3 shows why this occurs. Normally, even when a muscle is relaxed, there is sufficient stretch on the muscle spindle to cause some activity in the sensory fibers. However, when the muscle contracts and the tendons are brought closer together, the tension on the intrafusal fibers is reduced. This is known as unloading of the muscle spindle. Under physiologic conditions, unloading can be prevented by the activity of the gamma efferent system. When the gamma efferents are stimulated, they cause the contractile ends of the intrafusal fibers to shorten. As a result, the amount of tension of the cen-

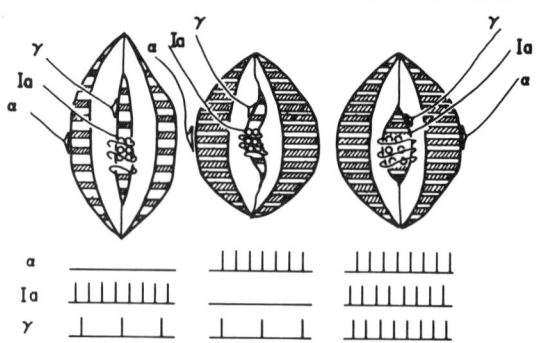

Fig. 13-3. Unloading of a muscle spindle occurs when the muscle contracts, because the tension on the central region of the intrafusal muscle fiber, where the Ia afferent terminates, is reduced. Unloading can be prevented by the gamma motoneuron, which shortens the intrafusal fiber in parallel with the extrafusal fiber.

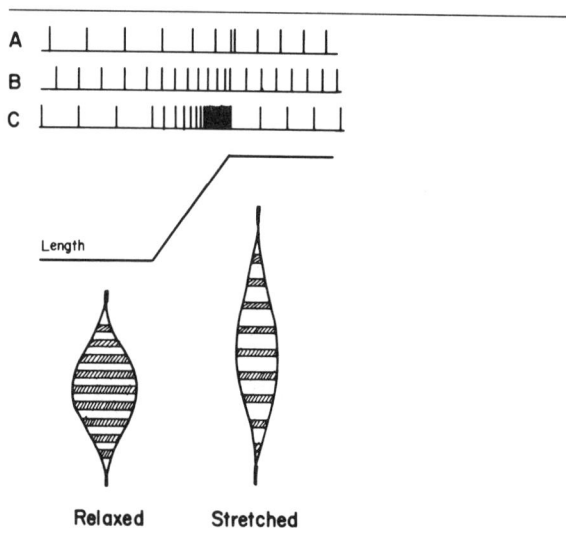

Fig. 13-4. The response to muscle stretch can be altered by stimulation of the gamma efferent fibers. The tonic firing rate is increased (B) over control (A) by the gamma static fibers without any change in the phasic firing rate. In contrast, gamma dynamic activity (C) causes an increase in the phasic response of the Ia afferent without having an effect on the tonic rate of firing.

tral region of the intrafusal fibers is maintained, and the afferent fibers continue to fire.

The two types of gamma efferent fibers have different effects on the muscle spindles (Fig. 13-4). Both of the gamma efferents cause an increase in the frequency of Ia afferent firing in response to a muscle stretch. However, the gamma dynamic fiber causes a large increase in the phasic response of the Ia fiber without having much of an effect on its tonic response. On the other hand, the gamma static fiber increases the magnitude sensitivity without having a noticeable effect on its phasic response. The differences in the effect of gamma dynamic and static fibers on the firing pattern of the Ia afferent are most likely due to their selective innervation of the intrafusal fibers (see Fig. 13-2).

The Ia afferent sends branches to both the nuclear chain and nuclear bag intrafusal fibers. The phasic response of the Ia afferent is caused by stretch of the nuclear bag fiber, while the static effect results from stretch of the nuclear chain fiber. Since the dynamic gamma efferent primarily innervates the nuclear bag fiber, it enhances the dynamic response to stretch. The increased responsiveness to the magnitude of the muscle stretch produced by the gamma static efferents is, analogously, due to its innervation of the nuclear chain intrafusal fibers. Another consequence of the selective innervation of the muscle spindle is that the secondary afferents (group II) exclusively innervate the nuclear chain fibers, the firing rate of which is affected only by the gamma static efferents.

THE WITHDRAWAL REFLEX

Spinal cord reflexes are the most elementary components of the motor control system. Through them, a number of simple motor acts can be performed in response to a stimulus without any input from the central nervous system. They also play a role in more complex motor behaviors initiated by the central nervous system. Finally, they serve as valuable clinical tools to test the integrity of both the peripheral and central nervous system.

A great deal of information about the organization of reflexes can be gained by studying the underlying neuronal circuitry of spinal cord reflexes. For example, the receptors responsible for initiating the withdrawal reflex are the pain receptors found on the terminals of small myelinated afferent fibers. After entering the spinal cord, these afferents give rise to a variety of neuronal circuits, some of which are illustrated in Figure 13-5.

The Withdrawal Reflex

The final common pathways for these circuits are the alpha motoneurons innervating flexor muscles. For example, if someone steps on a sharp tack, pain receptors in the foot are stimulated. These generate activity in pain fibers that in turn activate alpha motoneurons. The alpha motoneurons then stimulate flexor muscle fibers that

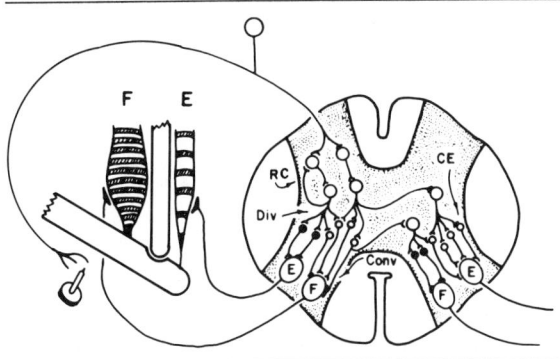

Fig. 13-5. Stimulation of pain fibers causes withdrawal of the foot from the painful stimulus. The incoming fibers diverge (Div) to stimulate interneurons, which are inhibitory to the extensor (E) motoneurons. Parallel pathways and recurrent collaterals (RC) maintain the withdrawal reflex even after the pain fiber stops firing. The many interneurons that converge (Conv) on the flexor (F) alpha motoneuron produce a vigorous response. Interneurons cross to the contralateral side of the cord to produce a crossed extensor (CE) reflex.

contract to move the foot away from the painful stimulus. The features that characterize the withdrawal reflex can be explained by the neuronal circuitry of the reflex within the spinal cord.

The pain fibers are small and so have a slow conduction velocity. Once the impulse initiated by the painful stimulus reaches the spinal cord, it must pass through several interneurons before reaching the flexor alpha motoneurons (making it a multisynaptic or polysynaptic reflex). This circuitry accounts for the delay that occurs between the stimulation of the pain fibers and the withdrawal of the foot.

Afterdischarge

Note also that Figure 13-5 illustrates a number of alternative pathways through which the afferent information can reach the flexor motoneurons, and that there are reverberating circuits in which an interneuron feeds back onto and reexcites a previously excited interneuron. These pathways cause the alpha motoneuron to be stimulated repeatedly so that the response outlasts the stimulus. This is called *afterdischarge* and accounts for the observation that the foot remains elevated even after it has been removed from the tack, giving the brain time to decide where on the ground the foot should be placed.

Local Sign

If the stimulus is not too great, as in the case of stepping on a sharp tack, then only the muscles controlling the stimulated area are activated. This phenomenon is called *local sign*. However, if the stimulus intensity is increased (as would occur if a hot coal were stepped on), then the muscles of the entire leg become involved. This phenomenon is called *irradiation*. Both of these phenomena result from the neuronal circuitry of the withdrawal reflex. Local sign reflects the fact that the greatest number of collaterals from the pain fiber synapse on alpha motoneurons going to the flexor muscles that underlie the area of stimulation. However, there are a sufficient number of collaterals going to the other leg muscles to activate them if the stimulus intensity is high enough.

Crossed Extension

Finally, in addition to activating motoneurons on the side stimulated (the ipsilateral side), the pain fibers cross to the other side of the spinal cord to activate contralateral extensor motoneurons. Thus, when the ipsilateral foot is withdrawn by the flexion reflex, the contralateral leg is extended to support the body. This phenomenon is called the *crossed extension reflex*.

Pain fibers are not the only neurons that give rise to polysynaptic flexor reflexes. Fibers originating in other cutaneous receptors and in joints, as well as the group II muscle afferents from the muscle spindles, all can cause flexion reflexes. All of these fibers have been grouped together and as such are called *flexor reflex afferents* (FRA). Their role in movement is not yet established, but they may be involved in aiding movement generated by the central nervous system.

For example, when a voluntary flexion is initiated by descending pathways, flexor motoneurons are activated. Once the movement begins, sensory fibers from the skin, joints, and muscles of the involved limb are stimulated and these, through the FRA pathways, add to the excitation of the spinal alpha motoneurons, thus reinforcing the voluntary movement. The interaction between reflex pathways and higher brain centers in initiating and controlling movement is extremely important, and many such examples are discussed in the chapters to follow.

MUSCLE REFLEXES
The Stretch Reflex

Compared to the flexion reflex, the circuitry underlying the stretch reflex is relatively simple. The sensory limb of the reflex is the Ia afferent neurons originating in the muscle spindles. The efferent limb is the alpha motoneurons innervating the muscles from which the Ia afferents originated. Within the spinal cord the Ia afferents synapse directly on the motoneurons, making this a monosynaptic (single synapse) reflex.

Because of the high conduction velocity of the afferent fibers and the single synapse interposed between the afferent and efferent limb of the reflex, the response latency is much shorter than it is for the withdrawal reflex. Since the only motoneurons stimulated are those going to muscles from which the Ia afferents originated, the reflex lacks the irradiation characteristic of the flexion reflex. Also, there are no reverberating pathways in the reflex pathway, so afterdischarge does not occur. Under physiological conditions, this reflex is initiated by stretching the muscle spindles; thus, it is called the *stretch reflex*.

The stretch reflex is an extremely important component of the motor control system and its characteristics must be well understood to appreciate how movements are coordinated. Before describing its physiological function, however, some additional features of its neuronal circuitry should be considered. As the Ia afferent enters the spinal cord, it branches to send collaterals to a number of neurons besides the alpha motoneurons. Several of these branches enter ascending spinal tracts to supply the central nervous system with information needed for it to monitor ongoing muscle activity.

Reciprocal Innervation

Another branch of the Ia afferent fiber synapses on spinal interneurons. These interneurons then form inhibitory synaptic connections, with the alpha motoneurons going to the muscles that are antagonistic to those from which the Ia afferent originated. This is called *reciprocal innervation* and is an extremely important organizing principle of the motor control system. In almost all cases, whenever a motoneuron is stimulated by an afferent neuron, the afferent sends a collateral to an interneuron that then inhibits the antagonistic motoneuron. Such an arrangement has the obvious advantage of allowing a movement to occur without interference from the antagonistic muscles.

Autogenic Inhibition

In addition to the monosynaptic stretch reflex from the muscle spindle, there is a disynaptic (two synapse) reflex originating in the Golgi tendon organ. The Ib afferent fibers from the Golgi tendon organ synapse on interneurons, which in turn form inhibitory synaptic connections with the alpha motoneurons returning to the muscle from which the reflex originated. This is called *autogenic inhibition*. Because the Golgi tendon organs are excited by muscle contraction, this reflex functions as an inhibitory reflex, reducing the amount of alpha motoneuron activity whenever a muscle contracts.

The Renshaw Cell

There is one additional neuronal circuit that is important to describe before discussing how spinal cord reflexes are used in the control of movement. Before the alpha motoneuron axon exits the spinal cord, it gives off a collateral that synapses on an inhibitory interneuron called the *Renshaw cell* (named for its discoverer). The Renshaw cell synapses with the motoneuron from which the collateral arose, as well as with other motoneurons going to the same and synergistic muscles. Its simplest function is to limit the rate of alpha motoneuron firing.

FUNCTIONAL SIGNIFICANCE OF SPINAL CORD REFLEXES

Having described the organization of the spinal cord reflexes, we can now proceed with a discussion of how they are used in the control of movement. Recall from the previous discussion that when a skeletal muscle is stretched, the intrafusal fibers of the muscle spindle are also stretched, eliciting a monosynaptic reflex that causes the motoneurons returning to that muscle to be activated. This reflex causes the stretched muscle to be contracted.

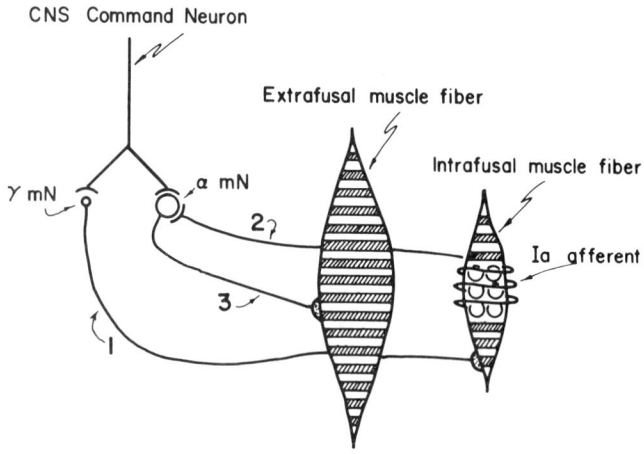

Fig. 13-6. *The gamma loop participates in the control of alpha motoneuron excitability. Central command neurons activate the gamma motoneurons (γ mN) (1), which increase the discharge in the Ia afferents (2), which then increase the firing rate of the alpha motoneurons (α mN) (3).*

The stretch reflex can be considered as a negative feedback system in which stretch on a muscle elicits a response that causes the muscle to counteract the stretch. This is the reflex evoked when a neurologist strikes the patella (knee) tendon with a reflex hammer. The tap on the tendon pulls on the muscle, activating the Ia afferent fibers from the muscle spindle. Although this is probably the most familiar way to activate the stretch reflex, it is not in any way analogous to how the reflex is used under physiological conditions.

The Gamma Loop

Figure 13-6 illustrates the circuitry of the gamma loop, which consists of the gamma motoneuron, the Ia afferent from the intrafusal fiber, and the alpha motoneuron. Theoretically, it is possible for the alpha motoneuron to be activated by a central command neuron acting through the gamma loop. If the command neuron descending from higher centers of the motor control system caused the gamma motoneuron to fire, the intrafusal muscle spindle would contract. This would generate activity in the Ia afferent, which would in turn cause the alpha motoneuron to fire. The alpha motoneuron would then complete the loop by activating its motor unit. In actuality, the gamma loop is not used to initiate activity, but it is important to the control of movement because it keeps the Ia afferents from unloading during a muscle contraction.

The Ia afferents serve two roles in the motor control system. One is to provide information about the length of the muscle to the central nervous system. The other is to contribute, along with the command neuron, to the excitation of the alpha motoneuron. Both of these functions would be lost if the Ia afferents stopped firing during a muscle contraction. Normally, this does not occur, because when the central command neurons fire the alpha motoneurons they also fire the gamma motoneurons. The coactivation of the alpha and gamma motoneurons allows the gamma loop to prevent unloading of the muscle spindle afferents during a muscle contraction.

Gamma Bias

The physiologic significance of gamma motoneuron discharge is demonstrated by the diagrams in Figure 13-7. Note that when the arm moves from a 45° to a 90° position, the command neuron increases the activity in both the alpha and gamma

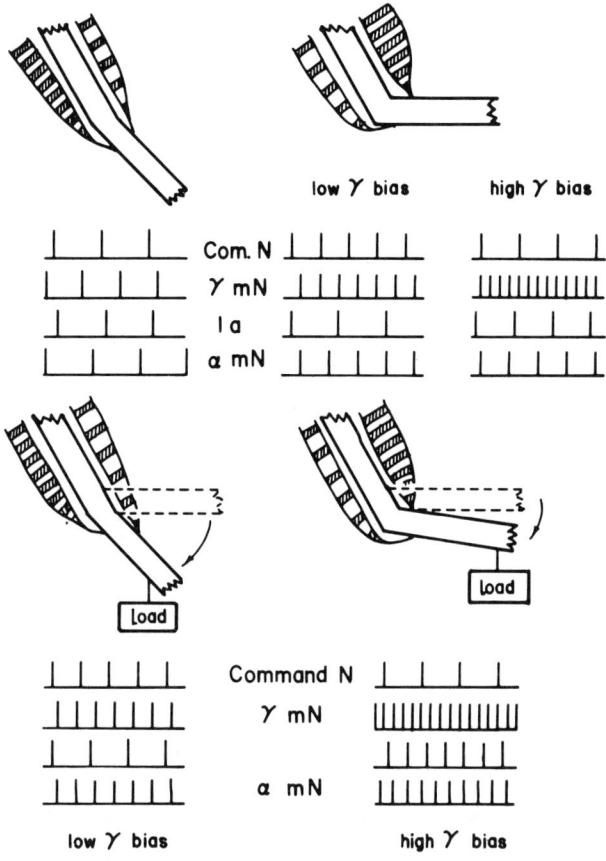

Fig. 13-7. How the gamma (γ) loop aids in the control of limb position. Compare the firing rate of the command neuron under conditions of low and high gamma bias. If gamma bias is high, an unexpected load does not cause the arm to fall as much, because the stretch reflex activation of the alpha motoneuron (α mN) is greater.

motoneurons. Two different conditions are illustrated. In one, labeled "low gamma bias," the amount of activity in the gamma motoneurons is minimal, and so little activity is generated in the Ia afferents by the gamma loop. In the other, labeled "high gamma bias", the amount of gamma activity is substantial, and the Ia afferent fibers contribute significantly to the excitation of the alpha motoneuron.

Functionally, these two states determine how much direct control the higher centers have over limb position. In the low gamma bias condition, most of the increase in alpha motoneuron activity is produced by the command neuron, and consequently it is in control of the limb. By contrast, in the high gamma bias condition the stretch reflex plays a much more important part in determining limb position.

One consequence of the amount of gamma bias is the different way the system responds when a sudden, unexpected load is placed on the limb (see Fig. 13-2). Note that the load causes a much smaller change in limb position under the high gamma bias condition. This phenomenon occurs because the reflex response to stretch is much more vigorous when there is a great deal of gamma motoneuron input to the intrafusal fibers. The greater Ia discharge in response to the stretch prevents the limb from moving as far as it does in the low gamma bias condition.

The type of gamma motoneuron activated also plays a role in determining how the limb responds to an unexpected load. If the predominant gamma bias were of the dynamic type, there would be a vigorous stretch reflex in response to a sudden change in load, but a slow alteration in load would have little effect on the stretch reflex. In contrast, a high static gamma bias would compensate for slow changes but would ignore sudden alterations in load.

These differences can have functional consequences. For example, in an activity such as skiing it is important to maintain an appropriate posture in order to keep your balance. The gamma loop system can aid the motor command neurons in holding that position. However, the stretch reflex cannot be too responsive to sudden changes in muscle length. Imagine what happens when a mogul is encountered. The least appropriate motor response is to resist the upward movement of the knees as the skis pass over the mound of snow. In this condition, static gamma motoneuron bias is required.

The Golgi Tendon Organ

The Golgi tendon organ also plays a role in informing the central nervous system about the status of the muscle and in regulating the amount of muscle contraction. Because it is in series with the muscle, the tendon organ is activated whenever the muscle contracts. Figure 13-8 contrasts the response of the Ia afferents to that of the Ib afferents from the Golgi tendon organ when a muscle contracts. Note that the frequency of firing in the Ib afferents is proportional to the amount of tension generated by the muscle. Thus, the Ib afferent is a tension detector, in contrast to the muscle spindle, which is a length detector.

Recall that the Golgi tendon organ is inhibito-

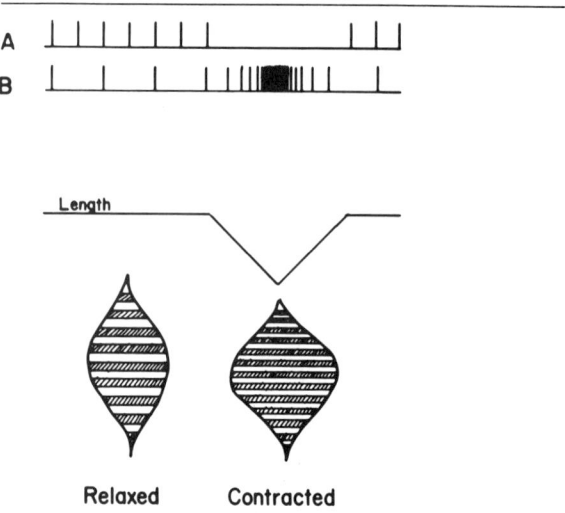

Fig. 13-8. When a muscle contracts, the Ia fiber (A.) decreases its activity, while the Ib fiber from the Golgi tendon organ (B.) increases its rate of firing.

ry to the alpha motoneuron so that when it increases its activity, it causes a reduction in the firing rate of the alpha motoneuron. Although the purpose of having a negative feedback system limiting the amount of muscle activation may not be immediately obvious, it can serve a useful purpose. Besides, the motor control system operates primarily by removal of inhibition rather than by direct excitation.

An example of how the Golgi tendon organ inhibitory reflex can aid in maintaining limb position is the situation that occurs when muscles begin to fatigue. Suppose you are trying to hold a book up in front of your eyes and your muscles begin to fatigue. The tension they generate decreases and your arm is unable to maintain its position. However, when the muscle generates less tension, the Ib afferent activity is decreased. This results in less inhibition of the alpha motoneuron. Consequently, the alpha motoneuron increases its firing rate and the force of muscle contraction is increased.

This reflex, like the stretch reflex, is able to compensate for changes in muscle performance without involving the central nervous system. Of course, the brain is informed of the changes taking place and a new strategy can be worked out; for example, the book can be put down. But in the meantime, the spinal cord reflexes have quickly and automatically maintained the original position set by the motor control system.

The Renshaw Cell

The Renshaw cell pathway is also a negative feedback system that limits the frequency of alpha motoneuron firing. Descending pathways of the motor control system can facilitate or depress the excitability of the Renshaw cells and thus control the frequency of alpha motoneuron firing. If Renshaw cell excitability is depressed, the amount of recurrent inhibition is reduced and the frequency of alpha motoneuron firing is increased. Using these pathways, the central nervous system can increase or decrease the ease with which spinal reflexes can elicit muscle contraction, or it can alter the effect that centrally generated motor commands have on spinal motoneurons. That is, if Renshaw cell activity is enhanced (and consequently, recurrent inhibition is reduced), then a small muscle stretch or weak central command is able to produce a powerful contraction.

Another, and perhaps more important, role for the Renshaw cell is enhancing motor contrast in the same way that horizontal cells increase contrast in the visual system. This phenomenon is possible because Renshaw cell axons have a much larger distribution to motoneurons going to synergistic muscles than to motoneurons going to homonymous muscles. That is, the muscles innervated by the alpha motoneuron whose collaterals activated the Renshaw cell (the homonymous muscles) do not receive as much inhibition as those alpha motoneurons going to the synergistic muscles. Thus, a generalized input to the spinal cord that excites both the muscle designated to make a movement and its synergists produces a much stronger contraction in the primary muscle, because it is less affected by Renshaw cell inhibition.

So far we have discussed a number of situations in which the built-in pathways of the spinal cord can be utilized by the central nervous system to change the influence that reflexes or centrally originating motor commands have on the skeletal musculature. The chapters that follow elaborate on this concept, and show that the built-in pathways are used extensively by the central nervous system to control and coordinate our movements.

14 The Brain Stem

The brain stem consists of the medulla, pons, midbrain, and parts of the diencephalon. It contains a number of neuronal circuits that are responsible for controlling such bodily functions as blood pressure, respiration, temperature, and feeding. In addition, it contains two important components of the motor control system, one of which is part of the reticular formation. The other consists of the vestibular apparatus and its associated brain stem nuclei. This chapter describes each of these systems and explains their roles in the control of movement.

THE RETICULAR FORMATION

All of the components of the motor control system within the brain, except for the pyramidal tract and vestibular apparatus, communicate with the spinal cord through the reticular formation. Two important descending pathways are involved. These are the medial reticulospinal tract, which originates from reticular nuclei in the pons, and the lateral reticulospinal tract, which originates from nuclei within the medial medullary reticular formation. The reticular nuclei from which these pathways originate receive their input from the cortex, cerebellum, basal ganglia, vestibular nuclei, and spinal cord. The reticular nuclei play a role in the execution of voluntary movements and in the coordination of posture and equilibrium.

Exactly how the reticular formation performs these functions is not well understood. However, the powerful influence that the reticular formation exerts on spinal cord reflexes can be revealed by experimental stimulation or lesions of the brain stem.

Spinal Shock

For example, if the brain of a cat is sectioned between the brain stem and spinal cord, it causes a condition known as spinal shock, in which there is a complete loss of reflex activity. Within a period of time (ranging from hours in the case of cats to months in the case of human beings) spinal neurons recover their excitability and all of the spinal reflexes described in the previous chapter can once again be elicited. However, because the modifying influences of the brain stem are lacking, the reflexes do not exhibit their normal behavior. For example, lightly touching the foot elicits, in addition to the expected withdrawal and crossed extension reflexes, a generalized contraction of all the flexor muscles. In many cases, the excitation spreads to the autonomic nervous system, producing a mass reflex in which flexion of the lower limb and trunk muscles is accompanied by evacuation of the bladder and rectum, and profuse sweating. As recovery proceeds, the excitability of the extensor alpha motoneurons increases to the point that chronic contraction of the extensor musculature results. All of these abnormal reflex effects are caused by the loss of the inhibitory influence of the reticular formation over spinal cord function.

The Motor Areas
of the Reticular Formation

The reticular formation contains both spinal excitatory and inhibitory areas. The inhibitory area is in the ventromedial portion of the medullary reticular formation. When this area is stimulated, it causes a profound inhibition of all extensor alpha motoneurons. Under normal circumstances, descending inputs from higher centers are required to maintain activity in the medullary inhibitory area. As a result, when the brain stem is sectioned above the reticular formation, neurons within the inhibitory area stop firing and there is an increase in the firing of alpha and gamma motoneurons to the antigravity muscles.

In contrast to the ventromedial portion of the medullary reticular formation, stimulation of the pontine reticular formation causes marked facilitation of the extensor alpha motoneurons. This area is under inhibitory control from higher centers, and as a result its activity is increased when the brain stem is sectioned above the pontine reticular formation. The medullary inhibitory area, while inhibiting extensor alpha motoneurons, also excites flexor alpha motoneurons via a reciprocal innervation pathway. Thus it is more correctly called an extensor inhibitory area. Similarly, the pontine excitatory area excites extensors and inhibits flexors and thus should be called the pontine extensor excitatory area.

Decerebrate Rigidity

Because of the organization of the brain stem, a midcollicular section of the brain stem in a cat does not produce spinal paralysis. Instead, there is a sudden increase in excitability of all the extensor reflexes and a decrease in excitability of all the flexor reflexes. The increased extensor alpha motoneuron excitability is so great that even without an external stimulus, the extensor muscles are kept in a state of steady contraction. This condition is called *decerebrate rigidity* although it resembles the clinical condition of spasticity, not rigidity. The reason for this becomes evident in the discussion of cerebral control of movement in Chapters 16 and 17.

The extensor tone within the legs of a quadrapedal animal, such as a cat, is so high that the animal can support itself in a standing position, with its back arched and its tail and head extended. These extensor muscles all support the animal against gravity and so are called *antigravity muscles*. In humans, lesions which sever the connections between the cerebrum and reticular formation also produce an increase in excitability of the antigravity muscles, but because of our bipedal standing only the legs are extended. The arms are held in a flexed position because in these limbs, the flexors and not the extensors resist the force of gravity.

The increased excitability of the antigravity muscles is caused by the spontaneous activity of neurons within the nuclei of the pontine excitatory reticular formation that give rise to the medial reticulospinal tract. Normally, these neurons are kept from firing at an excessive rate by descending inhibitory pathways from the cerebrum and cerebellum. When these are severed, the reticular formation is released from higher control and acts unopposed, causing increased tone in antigravity muscles.

Under physiological conditions, the medial reticulospinal tract (which is excitatory to antigravity muscles), and the lateral reticulospinal tract (which is inhibitory to antigravity muscles) act together to control normal movements. However, as indicated earlier, the lateral reticulospinal tract is not spontaneously active, so that when the brain stem is sectioned it is not able to oppose the action of the medial reticulospinal tract.

Excitation of antigravity muscles is brought about by two mechanisms. In one, the alpha motoneurons innervating them are directly excited by the descending pathways. In the other, the descending pathways indirectly excite the alpha motoneurons through the gamma loop. In most cases, both of these mechanisms are involved in producing the clinical signs resulting from lesions within the brain. But if one or the other of them is mainly responsible for the effect of the lesion, the term *alpha rigidity* or *gamma rigidity* is used to indicate the major cause of the motor deficit.

Release of Function

The phenomenon of release of function is important in understanding the clinical signs resulting from lesions within the central nervous system. Many of the components of the motor control system are responsible for producing a variety of motor behaviors. These components are normally kept under inhibitory control by higher centers of the motor control system. When needed to produce a movement, the inhibition is withdrawn and the motor behavior is automatically performed. If a lesion of the central nervous system causes these inhibitory pathways to be damaged, the lower control centers operate spontaneously to produce inappropriate or pathological motor activity. In this case, it is the pontine extensor excitatory reticular formation that is released from higher control and decerebrate rigidity results.

THE VESTIBULAR APPARATUS

The other major component of the motor control system found within the brain stem is the vestibular nuclei. These nuclei receive their primary sensory input from the vestibular apparatus, but also receive input from the cerebellum and from visual, somatic, and muscle receptors. The vestibular nuclei give rise to a number of vestibulospinal tracts, which do not differ sufficiently from a physiological point of view for us to distinguish between them. This section provides a brief description of the sensory organs within the vestibular apparatus and then discusses their role in the control of movement.

The vestibular apparatus consists of three semicircular canals, the saccule, and the utricle. These and the cochlea, with which the vestibular apparatus is continuous, are formed by hollow passageways within the temporal bone of the skull, called the *labyrinth*. The saccule connects to the cochlea which is below it and to the utricle which is located above it. The three semicircular canals emerge from the utricle.

The sensory cells of the vestibular apparatus, called *hair cells*, are contained within a membranous sac that is filled with endolymph (the same fluid that fills the cochlear duct of the ear). The space between the membranous sac and the temporal bone is filled with perilymph. An elaborate system exists within the semicircular canals, utricle, and saccule to activate the sensory cells.

The Semicircular Canals

The hair cells within the semicircular canals are located on mounds of epithelial tissue called the *cristae* (Fig. 14-1). The cristae are found within the ampulla, which is the widening of the semicircular canal where it joins the utricle. The hair cells are similar to those found within the cochlea. However, these hair cells are polarized (Fig. 14-2). That is, one of the hairs, called the *kinocil-*

ium, is larger than the 40 or so other hairs, which are called *stereocilia* (Fig. 14-2A). When the stereocilia are bent toward the kinocilium, the hair cell is depolarized (Fig. 14-2C). When they are bent away from the kinocilium, the hair cell is hyperpolarized (Fig. 14-2D).

The cilia extending from these cells are embedded in a gelatinous mass, called the *cupula*, that fills the lumen of the ampulla. When the head is rotated, the fluid within the semicircular canals lags behind the movement of the labyrinth and thus prevents the cupula from moving. As a result, the hairs embedded within the cupula are bent in the direction opposite to the direction in which the head is moving (Fig. 14-3).

The semicircular canals are oriented at right angles to each other. The horizontal canal lies in a plane that is approximately parallel to the earth when the head is held upright. The two vertical canals lie in planes perpendicular to the earth, with the anterior canal oriented along a line from the center of the head to the eye and the posterior canal oriented along a line from the center of the head toward the ear.

The semicircular canals work in pairs, so that when one is stimulated, the other is inhibited. For example, the hair cells within the left semicircular canal are polarized so that the kinocilium is on the side of the hair cell closest to the utricle (i.e., closest to the center of the head). As a result, when the head is rotated to the right, the stereocilia are bent away from the kinocilium and the hair cells are inhibited within the left semicircular canal (Fig. 14-3). The kinocilium in the right horizontal semicircular canal is also placed on the side of the hair cell closest to the center of the head. However, when the head rotates to the right, the endolymph within the right horizontal canal pushes the cupula toward the center of the head, stimulating the hair cells within the right semicircular canal.

A similar arrangement exists within the vertical canals. Since the anterior vertical canal is parallel to the posterior vertical canal on the other side of the head, it is these two canals that work togeth-

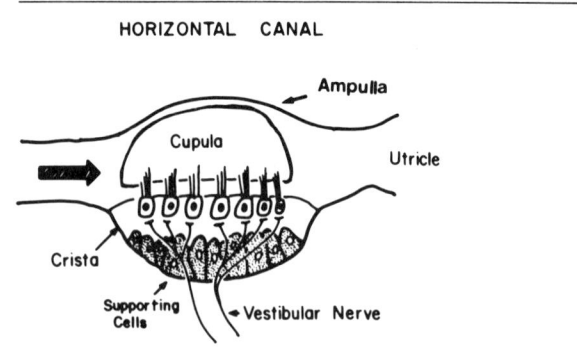

Fig. 14-1. The ampulla of the horizontal semicircular canal.

er. Movements of the head that cause the hair cells within one of the anterior vertical canals to be stimulated cause the hair cells in the contralateral posterior vertical canal to be inhibited.

If the rotation of the head continues, the elasticity of the cupula causes it to return to its original position and the stimulation or inhibition of the hair cell ceases. Thus, after approximately 20 to 30 seconds the activity of the semicircular canals returns to its resting levels and the perceptions and reflexes generated by the semicircular canals disappear. As a result, the semicircular canals only respond to changes in angular velocity (e.g., angular accelerations). Constant rotations, such as those produced by a twirling skater, do not produce long-lasting sensations. However, when the head ceases its movement, the endolymph continues to flow and as a result the cupula is pushed in the direction that the head was moving. Thus, at the conclusion of a rotation to the right, the hair cells on the right are inhibited and those on the left are stimulated. Again, after the head is still for about 20 to 30 seconds, the cupula returns to its resting position and hair cell activity returns to its normal resting value.

The Utricle and Saccule

The hair cells within the utricle and saccule are located on a mound of tissue called the *macula*. The cilia of these cells are embedded in a gelatin-

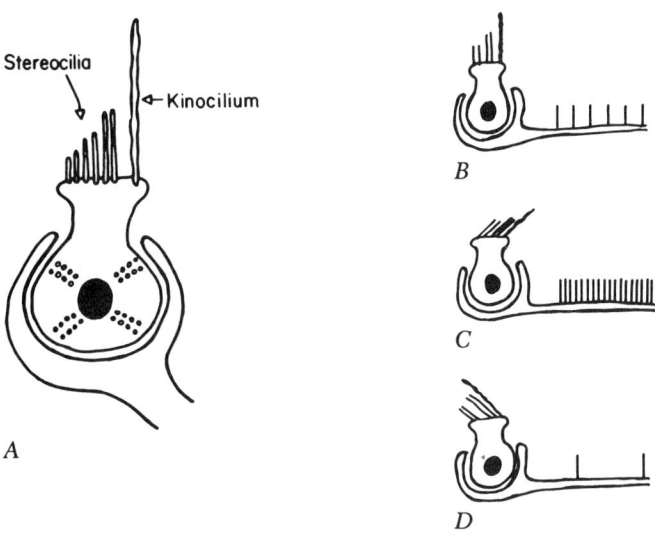

Fig. 14-2. A. The hair cells contain about 40 stereocilia and 1 kinocilium. B. At rest, the hair cells cause a steady discharge of the vestibular (VIII) nerve fiber innervating them. C. When the stereocilia are bent toward the kinocilium, the rate of firing increases. D. When the stereocilia are bent away from the kinocilium, the rate of firing decreases.

ous mass filled with crystals of calcium carbonate. The gelatinous mass is called the *otolith membrane* and the crystals are called *otoconia*. Because the crystals are heavier than the endolymph containing them, they tend to fall in a gravitational field and thus bend the cilia.

The macula of the utricle is in a horizontal position when the head is upright, while the macula of the saccule is in a vertical position. Like the hair cells of the semicircular canals, those of the utricle and saccule are polarized. The kinocilia are oriented so that some of the hair cells are stimulated and some inhibited, no matter how the head is oriented.

The hairs of these two organs (called *otolith organs*) respond at a steady rate as long as the head is kept in a steady position. They are tonic receptors whose firing rate provides the central nervous system with information about the head position. They also respond if the head is linearly accelerated. If an individual jumps off a step stool, he or she accelerates during the fall. The acceleration pushes the otoconia in an upward direction and thus stimulates those hair cells whose stereocilia are bent toward the kinocilium by upward movement.

Both the semicircular canals and the otolith organs provide the central nervous system with information about the position and movements of the head and participate in a variety of reflexes that maintain equilibrium during movement.

VESTIBULAR AND TONIC NECK REFLEXES

Like the reticulospinal tracts, the vestibulospinal tracts are excitatory to antigravity muscles and inhibitory to their antagonists. When the brain stem is sectioned above the pontine reticular formation, the neurons within the vestibular nuclei are also released from inhibition. As a result, they begin to fire continuously at a high rate. This activity, along with that of the excitatory pontine reticular formation, is responsible for producing decerebrate rigidity. The vestibulospinal tracts are essential to the production of decer-

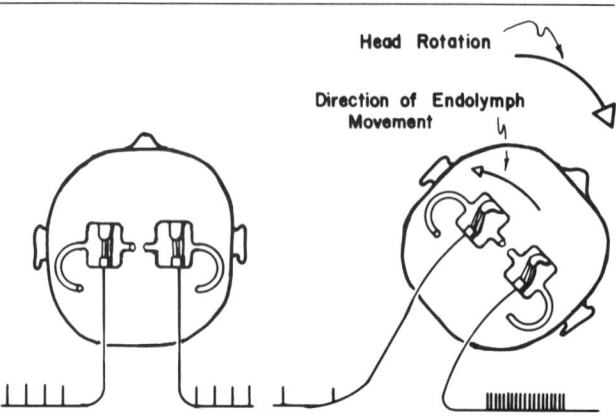

Fig. 14-3. *When the head is rotated to the right, the endolymph in the horizontal semicircular canals pushes the cupula toward the left. The stereocilia in the left horizontal canal are bent away from the kinocilium and decrease their rate of firing, while at the same time, the stereocilia in the right horizontal canals are bent toward the kinocilium and thus increase their rate of firing.*

ebrate rigidity because if the lateral vestibular (Deiters') nucleus is destroyed, the rigidity disappears.

Gamma and Alpha Rigidity

The rigidity produced by sectioning the brain stem is of the gamma type. The alpha motoneurons innervating the antigravity muscles are primarily excited through the gamma loop. This phenomenon can be demonstrated by cutting the dorsal roots and thus preventing Ia afferent fibers from reaching the alpha motoneurons. When this is done, the increased tone in the antigravity muscles is abolished.

In addition to the inhibition coming from the cerebrum, the Deiters' nucleus receives a powerful inhibitory pathway from the anterior cerebellum. This pathway is not cut when the brain stem is sectioned at the midcollicular level, and thus the neurons within the vestibular nuclei are still under some inhibitory control. If, along with sectioning the brain stem, the anterior cerebellum is lesioned, the spontaneous firing frequency of the vestibular neurons is increased substantially, and the rigidity of the antigravity muscles becomes even more exaggerated.

The rigidity produced under these conditions is of the alpha type, because when the dorsal roots are cut, the antigravity tone does not disappear. The dichotomy between alpha and gamma types of rigidity is somewhat artificial, in that lesions within the brain stem cause an increase in both alpha and gamma motoneurons. When the descending excitatory input is exceptionally high, as it is when all the inhibitory inputs to the reticular formation and vestibular nuclei are eliminated, the antigravity alpha motoneurons can be driven to fire even without the contribution of the Ia fibers. However, this doesn't necessarily mean that the input from the gamma loop to the alpha motoneurons is less important than that received directly from the descending fibers.

The contribution of the vestibular apparatus to the maintenance of tone in the antigravity muscles is not its only function. The vestibular apparatus also participates in a number of important reflexes that are responsible for maintaining an upright posture, keeping the head erect when the body is tilted, and coordinating the head and eyes so that the eyes can remain fixated on an object in the visual field when the head is turned.

The two components of the vestibular apparatus play different roles in the control of movement. The primary function of the semicircular canals is to adjust the position of the head and eyes in order to keep the direction of gaze constant during bodily movements, while the otoliths act to maintain the position of the head in relation to gravity. We shall consider the function of the semicircular canals first.

The Vestibuloocular Reflex

To see an object clearly, its image must remain focused on the foveae. If the head moves, there must be a compensating movement of the eyes to keep the image on the foveae. If the body moves, both the head and eyes must move to keep the direction of gaze constant. These movements are brought about by reflexes initiated within the semicircular canals.

When the head is turned or body position changed so that the head is rotated, the endolymph within the semicircular canals flows, causing the cupula to move in a direction opposite to the head movement. Movement of the cupula stimulates the hair cells within the ampulla, which in turn causes excitation of the axons within the vestibular (VIII cranial) nerve. These afferent fibers, through their synaptic connections within the brain stem, cause reflex contractions of neck and extraocular muscles to produce appropriate adjustments of the head and eye positions.

Nystagmus

The simplest way to demonstrate these reflexes is to seat someone in a rotating chair and observe the movements of his or her eyes. If the chair is moved to the right, then the eyes move to the left by an amount necessary to keep the direction of gaze constant. This response can also occur in darkness, indicating that visual input is not necessary for the reflex to function properly. If the rotation is continued far enough that the eyes reach the limit of their ability to move, they rapidly reverse direction, and look forward. This movement of the eyes is called *nystagmus*. In this case, rotation of the chair caused stimulation of the horizontal semicircular canal. Movements in other directions stimulate the other semicircular canals, producing vertical or rotary nystagmus.

When the rotation of the chair is suddenly stopped, the cupula is pushed in the direction of the previous movement and elicits reflexes that are opposite to those occurring during the initial rotation of the chair. The eyes now move slowly to the right and when they get as far as they can within their sockets, they move rapidly to the left. This is called *postrotatory nystagmus*. It continues until the cupula returns to its neutral position within the ampulla. This normally takes approximately 20 to 30 seconds.

Vestibulomotor Reflexes

During this time period, all of the other postural reflexes produced by the semicircular canals can be demonstrated. In the example cited above, the chair was rotated to the right and then suddenly stopped. This elicits the same set of vestibular reflexes as are generated when the head is suddenly moved to the left. For example, the eyes slowly deviate to the right.

In addition to the eye movements, there is also a spontaneous deviation of the head to the right. This too would aid in visual fixation if the head were actually moving to the left. Vestibular limb reflexes are also elicited. For example, if the individual sitting in the chair is told to raise and then lower his hand to the same position, his hand deviates to the right as if he were compensating for the fact that his body was turning to the left. Similarly, if he is asked to stand up and walk in a straight line, he tends to walk in a circle toward the right, again to compensate for the vestibularly induced sensation that his body is being moved to the left.

From a clinical point of view, these reflexes are an important sign of normal vestibular function. When the vestibular apparatus or its pathways within the brain stem are damaged, nystagmus may appear spontaneously or may not occur at all. For this reason, recording eye movements when an individual is at rest or during and after rotation is a very useful clinical tool. However, the clinical emphasis on nystagmus should not obscure the physiological role of the semicircular canals. After all, there are very few situations in which we are subjected to continuous rotation.

More often, the head is moved about in all directions during our normal activities, and vestibular reflexes serve to maintain visual fixation. For example, when driving a car there is a constant change in head position. If the eyes were to follow the head, visual images would be swept all over the retina and clear vision would be impossible. All three semicircular canals are stimulated by these movements, and their combined output contributes to the reflex adjustments that enable us to see clearly.

The Optokinetic Reflex

Visual fixation is not the sole responsibility of the semicircular canals; the visual system also plays a role. For example, after being spun around, post-rotatory nystagmus can be suppressed by visually fixating on an object in space. Dancers and ice skaters have learned to do this in order to continue their routines without the disturbing effects of the vestibular reflexes that were elicited by their previous twirls.

However, visual fixation cannot act as rapidly as the vestibular system. To prove this, try to fixate on your finger as you move it back and forth in front of you. Fixation is possible only if your finger is moving very slowly. Alternatively, if you now move your head from side to side and keep your finger still, your eyes move back and forth to maintain a clear image of your finger. This is true no matter how fast your head moves.

In the first case, visual information from the retina utilizes the so-called optokinetic reflex pathways to maintain the image on the foveae. Clearly, this reflex system does not function very well when the image is moved rapidly. In contrast, the vestibulooculomotor reflexes elicited when the head is turned are extremely efficient. One reason for the greater efficiency of the vestibular reflex system is that the semicircular canals detect acceleration of the head, and so can anticipate where the head is moving and initiate compensatory eye movements in advance of a change of position. The visual reflexes, on the other hand, can only respond after the image has moved away from the center of the fovea.

The Static Otolithic Reflexes

The other group of vestibular receptors are found within the utricle and saccule. These otolithic organs are responsible for producing the reflex adjustments that are necessary to maintain the proper orientation of the body and head in relation to the pull of gravity.

The effects of these reflexes can be described by considering the postural adjustments of a cat under a variety of circumstances. When a cat is walking down a flight of stairs, its head is dorsiflexed so that it is looking straight ahead. Alternatively, if the cat is perched with its forelegs on a table, its head is ventroflexed, again to keep the eyes looking forward. Similar reflex adjustments occur if the cat is tilted. For example, if a platform on which a cat is standing is slowly tilted toward the right, the limbs on the right side of the cat extend while those on the left flex, so that the body and head are kept in an upright position.

These reflex adjustments are initiated by the activity of the hair cells within the utricle and saccule. However, the static otolithic reflexes do not act in isolation. Their activity is modified by reflexes that are initiated when the neck is bent.

The Tonic Neck Reflexes

Tonic neck reflexes are elicited by sensory receptors found in the joints or muscles of the cervical spinal cord. Experimentally, it is difficult to separate them from those produced by the utricle and saccule. However, it appears that when the head is bent backward, the tonic neck reflexes cause the forelimbs to extend and the hindlimbs to flex. Thus, when a cat bends its neck backward to look at an object above its head, there is a reflex-induced postural adjustment that brings the body in line with the head, thus making it easier for the cat to maintain its gaze in the upward direction.

Another example of a tonic neck reflex aiding normal activity occurs when the cat turns its head to the right. The tonic neck reflexes that are evoked by twisting the neck toward the right cause an increase in extensor tone on the right and an increase in flexor tone on the left. When the head is turned to the right in preparation for a movement in that direction, the extension of the right leg induced by the tonic neck reflex provides a pivot for supporting the movement.

Neck reflexes also play a role in supporting the muscular activity associated with moving from a lying to a standing position. For example, when an animal lying with its left side on the ground attempts to stand, the first movement is a rotation of its head toward the left, bringing it into a vertical position. This is followed by an extension of the limbs on the left side to rotate the body so that it is oriented in the same direction as the head. The first movement, bringing the head to a vertical position, is facilitated by otolithic reflexes that are organized to keep the head vertical. The second movement, extending the limbs on the left side, is aided by the tonic neck reflexes.

The Dynamic Otolithic Reflexes

Because the otoconia are subject to the force of gravity, they also move under the influence of acceleration. Thus, if an individual suddenly moves forward the otoconia are pushed in the opposite direction, as if the individual were falling backward. To compensate for the apparent loss of upright position, the body is tilted forward. This reflex is thus useful in aiding the forward pitch in posture that is made when an individual begins to run. Similarly, when an individual is falling through the air, the otolithic organs are stimulated. They reflexively produce a contraction of the foot muscles that aids in the absorption of the forces produced when landing occurs.

These spinal and brain stem reflexes have all been described in terms of how they aid in the postural adjustments required for normal movements to occur. They illustrate the general theme that the motor control system is built so that lower centers can carry out ordinary movements and make adjustments for unexpected disturbances without the need for cortical input. The components of the motor control system within the basal ganglia and cerebellum are similarly organized.

15 Basal Ganglia

The previous chapters have described the roles played by the spinal cord and brain stem in the control of movement. These chapters have explained how the spinal cord, through its extensive reflex actions, is able to perform a variety of tasks without the active assistance of the motor cortex. If, during an intended movement, there is a sudden increase in the load that must be moved, spinal reflexes can bring about the increase in muscular force necessary to overcome the added resistance. Similarly, the brain stem has the neuronal circuitry necessary to maintain postural tone and to coordinate head and eye movements through the actions of the vestibular reflexes so visual fixation can be preserved during normal motor activity.

This chapter and the one that follows discuss the role of the basal ganglia and cerebellum in the control of movement. These structures are extremely important to the production of normal movement. However, the mechanism by which they perform their roles is not at all clear. Neither laboratory experiments nor clinical case studies have been successful in revealing how the basal ganglia and cerebellum work. However, these studies have furnished many clues from which some generalizations can be made. It appears that the primary function of the basal ganglia is to generate the motor commands that provide the postural support necessary for the smooth execution of a movement, while the cerebellum is responsible for generating the commands that control the force, duration, and sequence of muscular activity required to ensure that the movement is carried out in a coordinated manner.

As an example, consider the situation in which the arm is extended to pick up an object from a desk. The position of the legs, trunk, head, and shoulder must be adjusted so that the arm is free

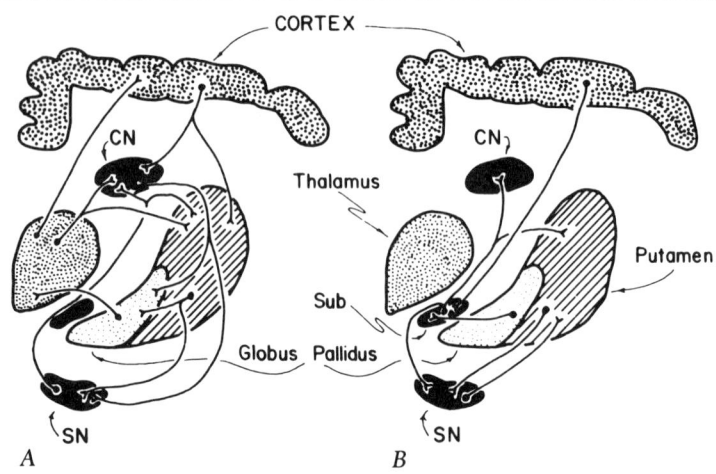

Fig. 15-1. Two major feedback loops within the basal ganglia. The basal ganglia consist of the caudate nucleus (CN), the putamen, the globus pallidus, the subthalamic nucleus (Sub), and the substantia nigra (SN). Damage to the pathway between the stratum and the substantia nigra causes parkinsonism.

to move toward the desk. These adjustments are carried out with the help of the basal ganglia. The muscles that extend the arm (the triceps, for example) must be contracted to initiate the movement, and the arm flexors (the biceps) must contract to slow the arm as it nears its target. Each of these muscles must be "turned on" at the right time and remain active for the proper duration if the hand is to reach the target efficiently. This task is carried out by the cerebellum.

The discussion that follows first describes the anatomical organization of the basal ganglia, then discusses the clinical signs of their dysfunction, and finally uses that information to explain their role in the control of movement. Although only the role of the basal ganglia in the control of movement is discussed, it is worthwhile to point out that the basal ganglia also have extensive connections with those areas of the brain concerned with cognitive and emotional activity. As a result, basal ganglia lesions produce behavioral deficits in addition to causing motor deficits.

THE ANATOMICAL CONNECTIONS OF THE BASAL GANGLIA

Figure 15-1 shows the cerebral nuclei that make up the basal ganglia. These are the *striatum* (composed of the *caudate nucleus* and *putamen*), the *globus pallidus* (or *pallidum*), the *subthalamic nucleus*, and the *substantia nigra*. Also included in the diagram are the major parts of the motor control system that communicate with the basal ganglia. Notice that there are no direct connections between the basal ganglia and the spinal cord or brain stem. This means that whatever roles the basal ganglia play in motor control are accomplished by modifying the motor commands of the cerebral cortex.

Input to the Basal Ganglia

The striatum is the major receiving area of the basal ganglia. Information reaches it from the cerebral cortex, the intralaminar nuclei of the thalamus, and the dopamine-containing neurons of the substantia nigra. In turn, the striatum sends its output to the globus pallidus and substantia nigra. Although the two parts of the striatum (the caudate nucleus and putamen) receive separate inputs and project to different portions of the globus pallidus and substantia nigra, there is not enough known about differences in their functional roles to warrant a detailed discussion of their individual pathways.

Output from the Basal Ganglia

The globus pallidus and substantia nigra are the major output centers of the basal ganglia. They send their projections to the thalamus, which in turn sends fibers to the cortex. A number of feedback loops are thus established. One, beginning within the globus pallidus, travels to the thalamus, the cortex, the striatum, and back to the globus pallidus. Another, which also starts in the globus pallidus and goes to the thalamus, reaches the striatum without a cortical loop. These pathways provide the globus pallidus with the ability to influence the output of the motor command neurons within the cerebral cortex.

Two other important pathways interconnecting the basal ganglia are also indicated in Figure 15-1. The subthalamic nucleus receives projections from the globus pallidus and motor cortex. It sends fibers back to the globus pallidus and substantia nigra and thus is in a position to influence all of the information leaving the basal ganglia. In addition, there is a feedback loop between the striatum and substantia nigra. The synaptic transmitter with which the fibers from the substantia nigra affect the cells of the striatum is dopamine. This is the well-known dopaminergic pathway that is damaged in Parkinson's disease.

It is important to realize that the pathways described here are mere outlines of the complex interconnections between the basal ganglia and other components of the motor control system. No attempt has been made to describe these pathways in any more detail, because doing so would not aid in understanding either the role played by the basal ganglia in producing movement or the deficits in motor function caused by basal ganglia lesions.

MOTOR DEFICITS PRODUCED BY BASAL GANGLIA LESIONS

Basal ganglia lesions have devastating effects on the motor control system, ranging from a loss of the ability to move (hypokinesia) to the presence of spontaneous movements that are impossible to control (hemiballismus). Because of the extensive interconnections between the basal ganglia and the close proximity of the basal ganglia to other cerebral structures, it has been difficult to assign exclusive responsiblity for a motor deficit to a particular basal ganglia nucleus. Nonetheless, a knowledge of the kinds of motor dysfunctions associated with basal ganglia lesions is very helpful in gaining an appreciation of their normal physiological role.

Deficits Due to Loss of Function

As indicated above, the globus pallidus is one of the major centers from which information leaves the basal ganglia. Thus, the effects of pallidal lesions, which remove a substantial part of the basal ganglia from the motor control system, should yield some insight into how movements are controlled by the basal ganglia.

The most striking effect of such a lesion is the loss of ability to maintain a stable postural position. For example, an individual suffering from a bilateral lesion of the globus pallidus is unable to keep the head or trunk upright. The head tends to fall over so that the chin touches the chest, and when walking the body bends at the waist so that the trunk assumes an almost horizontal position. This loss of postural stability is not due to muscular weakness or failure of voluntary control, because when asked to stand up straight these patients can do so easily. It thus appears that one of the important roles played by the basal ganglia is to provide the motor cortex with the information required to maintain an appropriate posture.

Deficits Due to Loss of Inhibition

In contrast to the effects of pallidal lesions, lesions elsewhere in the basal ganglia produce a release of excitatory activity, yielding such signs as rigidity, tremor, and spontaneous movements.

Lesions to the Subthalamic Nucleus

Damage of the subthalamic nucleus causes an involuntary flinging movement of the arms and legs on the side opposite the lesion, called *hemiballismus*. These movements are highly organized and appear similar to those made when an individual is suddenly thrown off balance. Although they can be held in check by intense voluntary effort, as soon as this effort is relaxed, the movements appear again. Surgical destruction of the globus pallidus or ventrolateral nucleus of the thalamus is often used to alleviate the symptoms of the disease.

These observations hint at a possible role for the feedback loops involving the subthalamic nucleus. Neuronal circuits within the striatum may be responsible for the organization of movements used to maintain balance when equilibrium is disturbed. These are presumably kept under control by the subthalamic nucleus. When required as part of a motor act, the subthalamic nucleus withdraws its inhibition and the motor commands are sent from the striatum to the globus pallidus and then on to the motor cortex. When the subthalamic nucleus is lesioned, the inhibitory pathways are removed, and the balancing movements occur spontaneously. If the pathway generating these movements is interrupted by destroying the globus pallidus or ventrolateral nucleus of the thalamus, then the ballistic movements can be reduced or eliminated.

Lesions to the Striatum

There are a number of clinical syndromes related to degenerative changes in the striatum, such as Huntington's chorea, which is characterized by ceaseless, uncontrollable activity of the limbs; athetosis, in which the limbs are continuously engaged in a slow, irregular, twisting motion; and dystonia, which is typified by twisting movements of the trunk and neck. All of these symptoms are release phenomena, indicating that an inhibitory pathway has been damaged. Although the actual pathways involved in these diseases are not know, it has recently been demonstrated that in Huntington's chorea there is damage to a GABAergic pathway (a pathway in which GABA acts as the synaptic transmitter) between the striatum and the substantia nigra.

Parkinson's Disease

The best understood of all the basal ganglia diseases is parkinsonism. This is a disorder caused by destruction of the dopaminergic pathway (a pathway in which dopamine is the synaptic transmitter) from the substantia nigra to the striatum. A variety of motor dysfunctions is observed in patients suffering from Parkinson's disease. The most common of these are rigidity, hypokinesia, and tremor.

Rigidity. The rigidity associated with Parkinson's disease is different from the spasticity associated with lesions of the descending cortical extrapyramidal motor pathways described in Chapter 14. Whereas spasticity results from an increase in tone of the antigravity muscles, rigidity is caused by an increase in muscle tone of all the muscles acting at a joint. It is often described as "lead pipe rigidity" because, although the limb offers resistance when moved by an examiner, it remains in the position to which it was moved. The resistance to passive movement is not constant. It alternately disappears and returns as the limb is being moved, giving rise to "cog wheel rigidity", another descriptive term.

Hypokinesia. The paucity of movement or *hypokinesia* observed in Parkinson's disease is not due to the loss of muscle power, since movements of normal strength can be made. Nor is the lack of movement caused by rigidity, since hypokinesia is occasionally seen without any sign of rigidity. However, those movements that are performed are carried out slowly and with a great deal of hesitation, and a variety of movements that normally do occur are absent. For example, the arms do not swing when walking and facial expressions do not change during a conversation.

Tremor. Physiological tremor, the spontaneous rhythmic oscillation of muscular force that accompanies all muscular activity, occurs with a frequency of 8 to 12 contractions per second. Its causes are not known, but it appears to be related both to the average frequency of alpha motoneuron discharge (which occurs at a rate of approximately 10 per second during most movements) and to the delay required for spinal cord reflexes to correct small perturbations in force. Pathologic tremors are divided into resting tremors (associated with Parkinson's disease) and intention tremors (associated with cerebellar disease). The resting tremor of Parkinson's disease occurs at a frequency of 4 to 7 per second and most of the time disappears during a voluntary movement. The characteristics of intention tremor are described in Chapter 16, The Cerebellum.

Each of the major signs of Parkinson's disease described above represents a release phenomenon caused by destruction of inhibitory neurons that originate in the substantia nigra. As indicated, these neurons use dopamine as a synaptic transmitter, and clinically it has been found possible to ameliorate some of the symptoms of Parkinson's disease by the administration of the dopamine precursor, L-dopa (dihydroxyphenylalanine).

Before the advent of L-dopa therapy, it was common to use anticholinergic agents to treat Parkinson's disease. Since acetylcholine is an excitatory transmitter used by interneurons within the striatum, it has been proposed that the feedback loop affected by Parkinson's disease consists of the nigral inhibitory neurons that project to the striatum, the cholinergic excitatory interneurons within the striatum, and a striatal inhibitory neuron that projects back to the stubstantia nigra. When the dopamine-containing neurons are damaged, the striatal output is released from inhibition. By increasing the concentration of brain dopamine, the striatum can be kept under control. The same result can be achieved, although not as well, by reducing the excitatory effect of the striatal cholinergic interneurons with anticholinergic drugs.

The actual pathway by which this feedback loop produces its effect on motor behavior is not known. Although these symptoms can be relieved by surgical lesion of the globus pallidus, indicating that the striatum is acting through the normal output channels of the basal ganglia, the nigrostriatal pathway is not the only one involved. Both surgical destruction of the globus pallidus and L-dopa therapy reduce rigidity much more than they reduce hypokinesia or resting tremor. It is thought that excessive activity within the limbic system also contributes to hypokinesia, and that resting tremor involves both basal ganglia and cerebellar feedback loops.

16 The Cerebellum

The cerebellum is an extremely important component of the motor control system. Not only is it involved in the coordination and control of all motor activity, it also receives information from all sensory systems and from every region of the cerebral cortex. The output of the cerebellum is sent to every part of the nervous system related to the control of movement. Removal of the cerebellum results in a profound disturbance of motor performance without affecting sensory awareness, intellectual capacity, or the ability to initiate movement. The discussion that follows first presents an overview of the neuronal organization of the cerebellum and then describes the role of the cerebellum in the control of movement, along with the deficits in motor behavior resulting from lesions within the cerebellum.

ANATOMY OF THE CEREBELLUM

From an anatomical viewpoint, the cerebellum is composed of an outer cortical layer and a group of nuclei that lie beneath the cortex. The cerebellar cortex is divided by two deep transverse fissures into three lobes: anterior, posterior, and flocculonodular (Fig. 16-1). It is further divided, by a series of shallow fissures, into ten lobules that are named in Figure 16-1.

The three cerebellar lobes developed at different times during evolution and perform different functional roles. The oldest part of the cerebellum, from a phylogenetic point of view, is the *flocculonodular lobe*. Because of its early development, it is also called the *archicerebellum*. This lobe is involved in the coordination of vestibular reflexes and consequently is also referred to as the *vestibulocerebellum*.

Next to develop evolutionarily was the *anterior lobe* or *paleocerebellum*, followed by the *posterior lobe* or *neocerebellum*. The correspondence

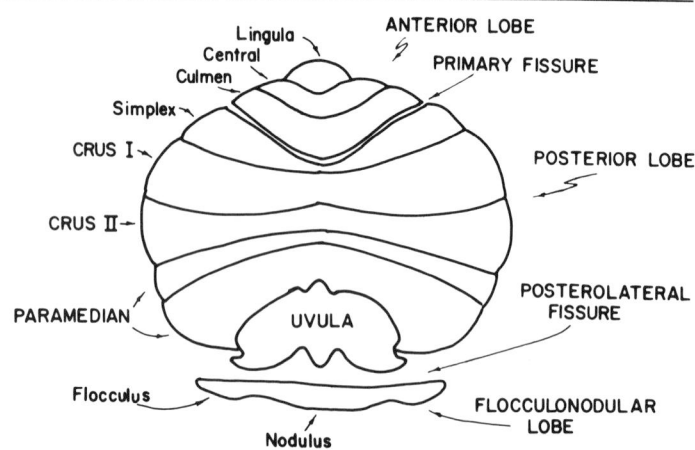

Fig. 16-1. *The cerebellar surface is divided into 10 lobules by a series of fissures.*

between the phylogenetic and functional divisions of the cerebellum is not exact. For example, the simplex and paramedian lobules (Fig. 16-1), although part of the posterior lobe, are operationally more related to the anterior lobe. This relationship has given rise to a more functional division of the cerebellum, based on the interconnections between the cerebellum and other parts of the nervous system.

The anterior lobe and those portions of the posterior lobe that receive information from the spinal cord are grouped together functionally as the *spinocerebellum*. Projections from receptors in the skin, joints, and muscles of a region of the body, as well as descending fibers from the motor cortex controlling the muscles in that region, all converge on a common area of the cerebellum. In other words, those regions receiving information from the motor cortex of the cerebrum also receive information from the peripheral receptors and thus are in a position to "know" both the "intent" of the cortex issuing a motor command and also, through sensory feedback, the results produced by that command. This gives the spinocerebellum the information it needs to coordinate muscular activity.

The lateral portions of the posterior lobe do not communicate with the spinal cord directly. Their major input is from the cerebral cortex and so can be considered as the cerebrocerebellum. However, unlike the spinocerebellum, which receives its cortical input from the motor cortex, the cerebrocerebellum derives its input from the association areas of the cortex (those parts of the cortex concerned with the generation of the idea for a movement). The output of the cerebrocerebellum is sent to the motor cortex and thus is in a position to control the pattern of activity responsible for a movement before a motor command is issued.

Output from the Cerebellum

The neuronal organization of the cerebellar cortex is the same in all of its regions. The most conspicuous neuron within the cortex is the Purkinje cell. This is the only neuron whose axon leaves the cerebellum, and thus it acts as the final common pathway along which the output of the cerebellum is directed. Interestingly and importantly, the Purkinje cell is inhibitory, which means that the cerebellar cortex functions by varying the amount of inhibition it produces on its target cells within the cerebellar and vestibular nuclei.

Purkinje cell axons from all regions of the cerebellar cortex, except the flocculonodular lobe, project to deep cerebellar nuclei (Fig. 16-2). Fibers from the anterior lobe (and other parts of

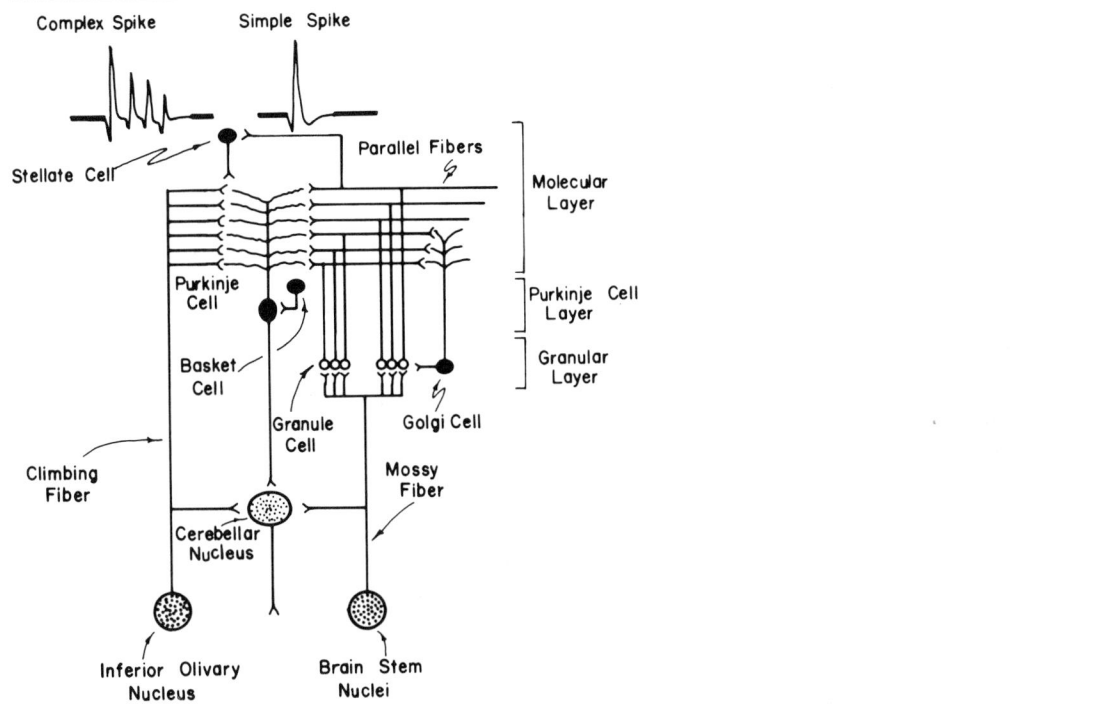

Fig. 16-2. The interconnection of neurons illustrated in the figure is repeated in all areas of the cerebellar cortex.

the spinocerebellum) project to the fastigial and interposed (globose and emboliform) nuclei, and those from the posterior lobe terminate in the dentate nucleus. The fibers from the flocculonodular lobe go to the vestibular nuclei and thus can be considered as functionally analogous to the cerebellar nuclei.

Inputs to the Cerebellum

There are two major input pathways to the cerebellum (Fig. 16-2). One is from the inferior olive of the medulla. Its axons are called *climbing fibers*. These axons, after entering the cerebellum, send one branch to the cerebellar nuclei and another to the cerebellar cortex. Within the cortex, they climb along the Purkinje cell, making multiple synapses on its dendritic tree. Each climbing fiber establishes synaptic contact with approximately 10 Purkinje cells. Discharge of the climbing fiber causes a single complex action potential to be generated in the Purkinje cell. The role of these complex action potentials in cerebellar function is not known.

The other major input to the cerebellum is derived from *mossy fibers* (Fig. 16-2). Each mossy fiber, after sending a branch to the cerebellar nuclei, terminates on approximately 50 granule cells contained in a cellular layer, called the *granular cell layer*, located just beneath the layer of Purkinje cells. Granule cell axons, called *parallel fibers*, travel to the superficial layer of the cerebellar cortex (the molecular layer) and then send two branches, approximately 1 to 2 mm long, in opposite directions. These branches synapse with the dendrites of approximately 500 Purkinje cells. The number of granule cells in the cerebellum is enormous, probably exceeding ten billion. Each Purkinje cell receives synapses from several hundred thousand parallel fibers. Mossy fiber activity results in a high frequency discharge of the Purkinje cells, and the cerebellar cortex is thought to exert its control on movement by way of this high-frequency discharge.

In contrast to the output of the cerebellum, all of the input fibers are excitatory. Thus, whenever an afferent pathway to the cerebellum is active, it causes excitation of the cerebellar nuclear cell and then excitation of the Purkinje cell in rapid succession. The Purkinje cell then causes inhibition of the nuclear cell, immediately terminating its output. Thus, afferent input to the cerebellum produces only short bursts of activity. Finally, the Purkinje cell is turned off by cerebellar interneurons.

These interneurons, the Golgi cell, the stellate cell, and the basket cell, are shown in Figure 16-2. They are all inhibitory neurons and are activated by the same incoming afferent fibers that excite the Purkinje cells. As indicated earlier, they cause the Purkinje cell to cease firing soon after it has been excited by incoming neuronal activity.

In summary, afferent fibers entering the cerebellum excite nuclear cells, Purkinje cells, and cerebellar interneurons in turn. This results in a burst of activity in the nuclear cells that is quickly inhibited by the Purkinje cells. Purkinje cell firing is then turned off by the inhibitory interneurons, freeing the circuit to respond to the next afferent input.

EFFECTS OF CEREBELLAR LESIONS

Despite the fact that the neuronal circuitry of the cerebellum is well understood, no clear view of how the cerebellum performs its role in the motor control system has emerged. However, by studying the effects of pathologic cerebellar lesions in human patients and experimentally-induced lesions in animals, it is possible to make some suggestions about how the cerebellum might work. The following paragraphs describe the motor deficits resulting from lesions to the three functional divisions of the cerebellum, in order to elucidate how each of them performs its particular role in the control of movement.

Lesions to the Flocculonodular Lobe

Lesions within the flocculonodular lobe (the vestibulocerebellum) cause deficits in vestibular function resulting in a loss of equilibrium and an ataxic gait. When standing an individual with such a lesion tends to fall forward or backward; when walking, his legs are spread far apart and staggers from one place to another. Also, this type of lesion can cause a spontaneous horizontal nystagmus to develop in which the eyes deviate slowly away from the side of the lesion.

The loss of vestibular function with cerebellar lesions indicates the importance of the cerebellum in coordinating activity generated in other neural centers of the motor control system. When the inhibitory influence of the Purkinje cells of the flocculonodular lobe is removed by the lesion, the vestibular nucleus on that side becomes overactive, causing the slow movement of the eyes characteristic of nystagmus. A curious aspect of damage to the flocculonodular lobe (particularly to the nodulus) is that motion sickness disappears.

Lesions to the Spinocerebellum

In cats, lesions of the spinocerebellum cause an increase in activity of the antigravity muscles (recall the discussion of muscle tone in Chapter 14, The Brain Stem). Since the cerebellar cortex is topographically organized, the muscles involved depend on the region of the anterior cerebellum that is damaged.

If the lesion is placed in the culmen (see Fig. 16-1), the forelimb is involved, while lesions in the centralis involve the hindlimb. Unilateral lesions cause an increase in tone of the extensor muscle on the side of the lesion and an increase in tone of the flexor muscles on the opposite side. These results indicate that the cerebellum plays a role in modulating the tone of the antigravity muscles and thus aids in coordinating the reflexes that maintain an upright posture. In primates, including humans, anterior cerebellar lesions rarely produce an increase in tone, most likely because of the great development of the cerebrocerebellum in primates.

Lesions to the Cerebrocerebellum

Lesions of the cerebrocerebellum in animals produce remarkably few deficits in motor performance, and those that do appear are easily compensated for. Similarly, in humans, surgical removal of tumors in the lateral cerebellar cortex does not result in any obvious clinical disturbances of motor performance. However, major disturbances in motor control result from lesions to the dentate nucleus (the target of Purkinje cells within the cerebrocerebellum). This suggests that the lack of deficits resulting from cortical lesions is probably related to the small areas involved and to the ability of intact areas to compensate for those parts of the cortex that were injured.

Most of the problems in motor performance produced by dentate nuclear lesions are also present in cerebellar disease, and thus can be used as a basis for speculation about the function of the cerebellum in the motor control system. These problems are all related to the loss of ability to produce smooth movements of the limbs from one point in space to another. Such a deficit is called *ataxia*.

Ataxia

An individual with cerebellar disease who is reaching for an object finds it difficult to move the other arm directly to the target. Instead of efficiently using all the muscles of the arm in their proper sequence, the movement is broken down into its constituent parts, with each joint moving separately. The arm first extends at the shoulder, followed by the forearm, and ultimately the hand. In addition, the beginning of the movement is delayed and the arm moves more slowly than normal. Even with the reduced velocity, there still is difficulty in stopping the movement, so that the arm most likely misses the object. Finally, all these movements are accompanied by a tremor that becomes worse as the hand gets closer to the object. Unlike the tremor in Parkinson's disease, which is present at rest, this tremor only occurs during movement and is called an *intention tremor*.

Adiodochokinesia

Another disturbance related to lesions of the cerebrocerebellum is loss of the ability to produce rapidly alternating movements, such as turning the hand back and forth. This deficit is called *adiodochokinesia* and occurs because each movement is first delayed in starting and then carried out for too long a time.

Based on the preceding summary of deficits in motor behavior caused by cerebellar lesions, it appears that the role of the cerebellum is to coordinate the sequence and duration of muscle activity in the performance of a movement. The interconnections between the cerebellum and cortex permit the cerebellum to participate in the planning of a movement, while the feedback from sensory receptors in muscles, joints, and skin provides the cerebellum with the information necessary for it to alter the timing of muscular activity during the execution of a movement.

17 The Cerebral Cortex

THE MOTOR CORTEX

In general terms, the motor cortex includes all those regions of the cortex that produce muscle activity when electrically stimulated. Defined this way, a large area of the cortex (Fig. 17-1) can be considered as "motor". The cortex is divided into a number of functional components that are concerned with the execution of voluntary movements. These are the *primary motor area*, which occupies the precentral gyrus (corresponding to the cytoarchitectural area 4), the *secondary motor area*, which is located in front of the primary motor area (area 6), and the *supplementary motor area*, which is located on the medial surface of the cortex (also area 6). In addition to these areas, other regions of the cortex not specifically concerned with the generation of voluntary movements, such as parts of the prefrontal and sensory cortex, also produce movements when stimulated and so are included in the definition of the motor cortex. Finally, there are cortical areas that do not produce any movements when stimulated but are nonetheless integral parts of the motor control system. These areas, in the frontal and parietal lobes, are termed the *association cortex* and are involved in the integration of cortical information and the elaboration of the "idea" required for the initiation of a voluntary movement.

The Motor Areas of the Cortex

The motor areas are organized in a topographic fashion similar to that seen in the sensory cortex (Fig. 17-1). Those muscle groups that produce fine movements have the largest cortical area devoted to them, while those that produce coarser movements, such as the trunk muscles, have little cortical area reserved for their control.

As can be seen in Figure 17-1, the secondary motor area projects only to the trunk and proximal muscles and thus its function appears to be

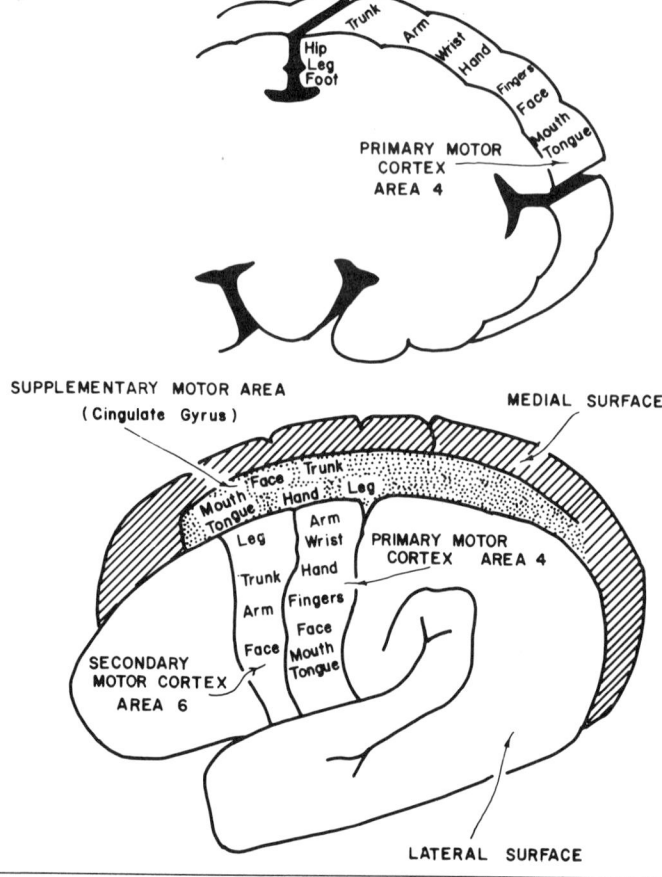

Fig. 17-1. The primary motor cortex has a precise topographic representation and is responsible for movements requiring fine motor control. The secondary and supplementary motor areas control global movements utilizing the more proximal muscles.

limited to the support of the body and limbs during movement. In contrast, the supplementary motor area has a complete representation of the body. When stimulated, it causes activity in more than one muscle, and so appears to control stereotypic movements involving several groups of muscles. However, the large size and extensive connections of the secondary and supplementary motor areas make it likely that these areas are involved in more than the production of the postural support required for muscular activity.

Lesions to these areas in humans provide some indication of the role they may play. For example, if the secondary motor area is damaged, there is a decrease in the ability to carry out motor activities requiring visual guidance. Because of this and the large number of fibers from the visual cortex reaching the secondary motor area, this area is thought to be involved in the performance of movements requiring visual feedback for their correct performance. In an analogous fashion, the supplementary motor area receives an extensive input from the somatosensory cortex and so may perform a similar task for sensory-guided motor behavior.

The Pyramidal System

The primary motor cortex is involved in the execution of fine movements using two different neuronal systems. These are the pyramidal and the extrapyramidal pathways.

From an anatomic point of view, the axons of the neurons making up the pyramidal pathway pass through the medullary pyramids on their way to the spinal cord. These neurons send collateral fibers to all of the subcortical components of the motor control system (the basal ganglia, cerebellum, vestibular nuclei, and reticular formation) discussed in the previous chapters. The information received by these centers from the motor cortex is used to aid in the production of motor behavior.

In addition to the fibers traveling from the cortex to the spinal cord (the corticospinal tract), there are fibers traveling from the primary motor area to the brain stem motor nuclei that are responsible for controlling the muscles used for facial expressions and speech. Although these neurons do not pass through the pyramids, since they do not go to the spinal cord, they serve the same function as the corticospinal tract and so, from a functional point of view, are included in the pyramidal system.

All of the neurons contributing to the pyramidal system do not come from the primary motor area. In primates, and presumably in humans as well, only about 30% of the neurons forming the pyramidal system come from area 4. Another 30% come from area 6 (the secondary and supplementary motor area), while the rest come from the parietal lobe (primarily from the somatosensory receiving area discussed in Chapter 7, Cutaneous Sensation).

The Extrapyramidal System

The extrapyramidal system consists of all those fibers originating in the cerebral cortex that are concerned with the production of movement but do not pass through the pyramids. Some confusion exists in the use of the term *extrapyramidal*, since in addition to being defined as we have done, it is also used to indicate the neuronal pathways originating in the basal ganglia. We will use it exclusively to refer to the nonpyramidal neurons originating in the motor cortex.

All of the extrapyramidal fibers terminate within the brain, ending on other components of the motor control system, along with collaterals from the pyramidal system. Its neurons communicate with the spinal cord through two brain stem pathways; these are the reticulospinal and rubrospinal tracts. These tracts are located in the dorsolateral positions of the spinal cord in close association with the cortico-spinal projections. All of these tracts terminate in the lateral gray area, which, as previously noted, contains alpha motoneurons innervating the more distal muscles concerned with the production of fine movements. Because of this anatomic organization, the extrapyramidal system can substitute for the pyramidal system if the latter is damaged. And, in fact, experiments with nonhuman primates and a number of clinical observations substantiate the view that, except for the finest movements requiring the greatest dexterity, the extrapyramidal system can govern the motor control system quite well. In contrast, the reticulospinal and vestibulospinal pathways concerned with postural adjustments descend in the ventromedial cord and innervate the more medial alpha motoneurons that control the trunk and proximal muscles. These are the tracts that produce the decerebrate rigidity that is associated with cortical and brain stem lesions.

CORTICAL CONTROL OF MOVEMENT
The Cortical Cell Layers

The neurons within the cerebral cortex are organized into six layers. The first three layers and the sixth layer are primarily concerned with integrating information within a particular area of the cortex or communicating with other cortical areas. Sensory information from the thalamus projects to the neurons in layer IV, while the major output fibers from the cortex originate in layer V. In the motor cortex, the fifth layer is particularly large because the neurons in this layer are used by the motor control system to modulate the activity of all the other components of the motor control system.

The neurons of layer V, called *pyramidal cells* because of their shape and not because of their contributions to the pyramidal tract, are among the largest cortical neurons. Particularly large pyramidal cells are found in the primary motor area (area 4); these are called *Betz cells*. The axons of these cells enter the pyramidal tract and terminate directly on motoneurons that supply muscles of the hand concerned with the performance of fine movements. There are only a small number of Betz cells (only 30,000 of the 1,000,000 cells making up a pyramidal tract are Betz cells), but because they form monosynaptic synapses on spinal alpha motoneurons, they are able to control the generation of small, precise movements. Other pyramidal cells also end directly on alpha motoneurons, particularly those that innervate the muscles of the arms and hands, but the overwhelming majority of pyramidal tract neurons synapse on interneurons within the ventral horn of the spinal cord and not on alpha motoneurons.

Cortical Cell Columns

Most of the pyramidal tract neurons controlling the activity of a spinal motoneuron pool are located in a region of the cortex about 1 mm in diameter. This is presumably true for the extrapyramidal system neurons as well, although studies to demonstrate this have not been done. These cortical columns, which contain neurons projecting to a specific motoneuron pool, have sharp boundaries but overlap with other cortical columns controlling different spinal motoneuron pools.

Spinal motoneuron pools innervating muscles acting at a single joint or performing a coordinated movement are influenced by overlapping cortical columns. Thus, the column responsible for activating a particular muscle is next to the column causing the inhibition of the antagonistic muscle. Similarly, the two columns that produce activation of two antagonists at a joint, and thus are capable of stabilizing the position of that joint, are also found next to each other.

Depending on the location and strength of the stimulus, a variety of coordinated movements can be produced by electrical stimulation of the motor cortex. For example, a limb can be made to extend with the appropriate amount of extensor excitation and flexor inhibition. These experimental results, taken together, indicate that there are regions of the cortex that act together to produce coordinated movements, even though the cortex is organized so that neurons exciting or inhibiting spinal motoneuron pools are grouped together in discrete columns.

Each of the cortical motor columns receives a rich supply of afferent information from somatosensory, muscle, and joint receptors. In general, the input to a particular column comes from the receptors that would be activated during a movement generated by that column. For example, the motor columns responsible for generating a grasping motion receive sensory input from the joints of the hand, the muscles used in closing the hand, and the receptors on the inner surface of the hand. This afferent supply is used by the cortex to correct movements during their execution in a manner similar to that described in the discussion of the gamma loop in Chapter 13, Spinal Cord Reflexes.

In addition to the stretch reflex occuring in the spinal cord, there is a reflex that involves the cerebral cortex; this is called the *functional stretch reflex*. The reflex pathway starts with peripheral receptors that transmit information through the lemniscal pathways to the thalamus and from the thalamus to the somatosensory receiving area in the postcentral gyrus of the cerebral cortex. From here, the information is transmitted to the primary motor area, completing the afferent limb of the reflex pathway. The efferent limb is carried along the pyramidal system to the spinal alpha motoneurons. This reflex aids the spinal cord in responding to changes in load imposed on a muscle during a movement.

That the primary motor cortex is responsible for generating motor activity is indicated by the observation that pyramidal tract neurons become active prior to the initiation of a movement. However, unlike the cells of the motor association cortex, these cells only respond if the movement is to be carried out (i.e., they do not signal an intention to move). Thus, in a movement such as reaching out to grasp an object, a group of pyramidal tract neurons discharge prior to the excitation of the alpha motoneurons, causing the sequence of muscle activity required for the smooth performance of the motor task. These motor cortex neurons are programmed by the association cortex to produce the pattern of muscular activity necessary for a coordinated movement to be performed. During the movement, sensory feedback from the periphery can be used to make whatever adjustments in cortical activity are necessary to compensate for a failure of the initial program to produce the requested movement.

Corollary Discharge

To show the contribution of corollary discharge to motor performance, studies have been performed on animals in which all sensory feedback had been surgically eliminated. At first, these animals made no spontaneous movements. In fact, it had been assumed that sectioning of the dorsal roots produced a paralysis as profound as that resulting from cutting the ventral roots. However, these animals soon learned to use their limbs and were able to perform motor tasks with almost as much skill as before the operation, demonstrating the ability of the motor control system to function without the benefit of any external feedback from the periphery. If these animals were blindfolded, so that they had no way of knowing what their limbs were doing or where in space they were, they could still learn to make a skilled movement such as squeezing a rubber bulb. The information used to acquire the conditioned response must have come from corollary discharge (internal feedback) circuits.

The corollary discharge circuits may also influence the conscious perception that is developed during a movement. For example, the nervous system must be able to distinguish between active and passive movements. Otherwise, there would be no way to know whether movement of an image across the visual field was due to the movement of the object or to the movement of the eyes. One possible way in which this discrimination could be made involves corollary discharge.

If the brain is made aware of the action that the muscles are taking (for example, whether the extraocular muscles are moving the eyes), it would be able to correctly interpret the sensory information it receives from the retina as an image passes over it. The confusion resulting from the lack of corollary discharge can be easily demonstrated in an experiment in which an eye is passively pushed from left to right. The result of this movement is that the image moves to the right. The perception reported by the subjects in this experiment is that objects in the visual field have moved to the right, not that the eye has been moved. The explanation offered to explain these results is that under normal circumstances the only way for an image to move over the retina when the eyes have not been actively moved is for the visual target to have moved. If these same subjects are asked to move their eye to the left but the movement is impeded by the experimenter, the visual perception reported is that the visual target moved to the left. This phenomenon is explained by noting that under normal circumstances if a command to move the eye is given, the eye moves, and if no movement of the retinal image occurs, then the object must have moved along with the eye; that is, the object must also have moved to the left. These experiments are interpreted to mean that the motor commands issued by the cortex are used by the brain to generate conscious perceptions.

IV Cerebral Physiology

18 Sleep

The general level of physiological activity varies periodically during the day. For example, body temperature is about 1°C higher in the evening than it is during the early morning hours, and adrenocortical hormones are secreted in greater amounts in the morning than at night. These and other biological rhythms that fluctuate over a twenty-four hour period are called *diurnal or circadian rhythms*. They are synchronized to the day-night cycle by the variation in levels of illumination that occur as the earth turns on its axis. If an individual is isolated from environmental stimuli, an internal biological clock is still capable of producing a cyclic variation in physiological activity. In this case, however, the cycle length is approximately twenty-five hours. It is necessary to be aware of these fluctuations in activity during the day when assessing clinical function. For example, in following the course of a fever, temperature should be taken at the same time each day.

The most obvious of the circadian rhythms is the sleep-wake cycle. During the day most individuals are awake and able to perform those activities, such as eating, drinking, learning, and procreating, that are necessary for individual and species survival. However, at some point within a twenty-four hour period, wakefulness becomes difficult to sustain and an individual passes into a sleep state. During this state consciousness is lost and most physiological functions decrease their activity.

Physiologic Role of Sleep
The biologic role played by the sleep state is not understood. It appears, from an evolutionary point of view, as early as the reptiles and is a dominant form of behavior in birds and mammals. Its irresistable nature, despite our best efforts to remain awake, and the severe behavioral disorders that result from a pathologic inability

to sleep, attest to its importance for normal function. One possible reason for sleep to have evolved is to keep animals immobile and safe from predators except during those brief periods when they need to perform essential biologic functions. Another is to provide time for the body to replenish energy stores that may have been depleted, or to repair tissue that may have been damaged during periods of stress. Still another reason is that a period of sleep is necessary for the brain to consolidate all that it has learned during the day and to place it in long-term memory.

The states of sleep and wakefulness can be assessed most objectively by electrically monitoring an individual's brain activity. This can easily be accomplished with electrodes placed on the scalp. The electrical activity recorded is referred to as an *electroencephalographic recording* or an EEG.

THE ELECTROENCEPHALOGRAM

Recording the EEG is a standard part of the neurological examination whenever brain dysfunction or epilepsy is suspected. Sixteen electrodes are usually used, eight on a side, and are placed so as to monitor the electrical activity from the entire cerebral and cerebellar cortex. The electrical activity recorded by an EEG electrode reflects the synchronized variation in the membrane potentials of millions of cortical cells. These fluctuations in membrane potential are caused by the EPSPs and IPSPs produced by afferent fibers ascending from the thalamus.

During an EEG examination, the frequency and the amplitude of the EEG waves are measured and the presence of any unusual waveforms noted. Focal lesions within the brain can be identified by the asymmetry in the recordings from the two sides of the brain. More diffuse lesions are indicated by large-amplitude, low-frequency brain waves. The most useful diagnostic value of the EEG is in the identification of epilepsy. During an epileptic seizure, which may or may not have any behavioral correlates, the EEG

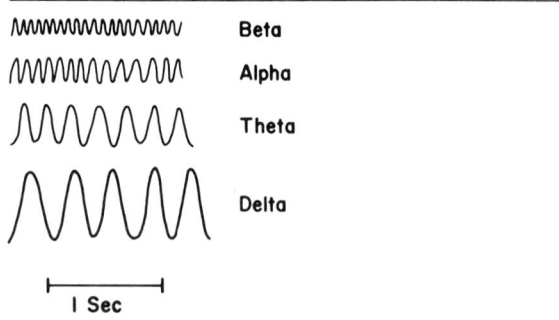

Fig. 18-1. As an individual progresses from the waking state to the deep stages of sleep, the brain waves become lower in frequency and higher in amplitude. Beta waves characterize the waking state (and the REM stage of sleep), alpha waves characterize the resting state, and the theta and delta waves are characteristic of the deep stages of slow wave sleep.

wave pattern becomes grossly abnormal, displaying a high-frequency discharge that is periodically interrupted by large spikes. More recently, with the advent of organ transplants, the absence of an EEG wave has been used as one of the major criteria upon which the clinical diagnosis of death is made.

The normal EEG pattern is generated by fibers arising in the thalamus. The neurons from which these fibers originate are controlled by the reticular formation. Thus, it is the reticular formation that is responsible for the variation in frequency and amplitude that occurs in the EEG waves as the state of alertness changes.

EEG waves are classified by their basic frequency (Fig. 18-1). The slowest rhythm, having a frequency of 1 to 3 Hz, is called the *delta* rhythm. EEG waves with a frequency of 4 to 7 Hz are called *theta waves* and those with a basic frequency of 8 to 13 Hz are called *alpha waves*. If the frequency of the EEG waves is above 13 Hz, they are referred to as *beta waves*. In general, the higher the frequency, the smaller the amplitude of the waves. Characteristic changes occur in the frequency and amplitude of the EEG during the sleep-wake cycle.

THE STATE OF SLEEP

When an individual is awake, he or she is alert and able to respond to changes in environmental stimuli. This capacity is lost when the individual enters the sleep state. When asleep, an individual is generally quiet and unaware of his or her physical surroundings. Physiological functions such as heart rate, respiration, blood pressure, and muscle tone all decrease. However, sleep is not simply a formless loss of consciousness. During the sleep period, the depth and type of sleep change in a predictable and cyclic manner. These changes in the state of sleep are correlated with unique behavioral and EEG characteristics.

The Alerting Response

The EEG recorded from an individual sitting quietly is characterized by an alpha rhythm. Muscle tone is low, but sufficient to keep the body in an upright position. Heart rate, blood pressure, and respiratory rate all are slower than normal. The eyes may be either opened or closed, but the individual does not respond to changes in the environment. However, if the individual is disturbed, the eyes immediately open, postural tone and autonomic activity increase, and the EEG pattern changes from an alpha to a beta rhythm.

This response is called the *alerting response* because of the sudden increase in awareness, or *alpha blocking* due to the change from an alpha to a beta EEG rhythm. A similar response occurs if the individual is asked to solve a problem or pay attention to a stimulus. The response is mediated by the *ascending reticular activating system* (ARAS). This is a major pathway ascending from the reticular formation to the thalamus through the core of the brain stem and is responsible for maintaining wakefulness.

The Stages of Slow Wave Sleep

If the individual remains in the alpha state for a period of time, there is a gradual decrease in consciousness that is characterized by a change in EEG activity. Instead of a continuous record of alpha waves, periodic episodes of large amplitude waves with a frequency of about 14 to 17 Hz are observed in the EEG. These are called *sleep spindles* and mark the initiation of the sleep state.

Over time, the depth of sleep changes. During the initial period of sleep, which is characterized by the sleep spindles, it is easy to arouse the individual. However, as time progresses, the state of sleep becomes deeper and deeper and arousal becomes more difficult to produce. As the depth of sleep becomes greater, the EEG waves decrease in frequency and increase in amplitude, autonomic activity decreases, and muscle tone decreases markedly.

The depth of sleep has been divided into levels I–IV. During stage I sleep, the dominant EEG pattern is of alpha waves and sleep spindles. As sleep becomes deeper (stages II and III), more and more of the EEG waves are of the theta type. Finally, as the deepest stage of sleep is reached (stage IV), delta waves characterize the EEG record. This type of sleep is called *slow wave sleep* (SWS) because it is characterized by slow EEG waves.

REM Sleep

There is another, quite different state of sleep, in which the EEG displays a preponderance of beta waves. This state of sleep is called *rapid eye movement* (REM) sleep (because of the rapid eye movements that accompany it), or *paradoxical sleep* (because of the apparent dissociation of behavioral and EEG signs). Dreaming occurs during REM sleep, so this type of sleep has also been called *dream sleep*. Slow wave and rapid eye movement sleep are quite different from each other. In addition to the difference in the EEG patterns and the presence of rapid eye movements and dreams in REM sleep, there are other neurophysiological differences. For example, in SWS muscle tone is decreased but not absent. In REM sleep there are periodic episodes during which all motor tone disappears. Although temperature falls in SWS, this is due to a fall in the set point at which body temperature is regulated. In contrast, during REM sleep thermoregulatory reactions, such as vasodilation in response to increased temperature or shivering in response to a fall in temperature, do not occur. Finally, there is a difference in the degree of alertness present after being aroused from SWS and REM sleep. When awakened from SWS, an individual is often groggy, confused, and unaware of his or her surroundings. The situation is quite different when awakened from REM sleep. In this case, the person is immediately alert and in control of his or her intellectual functions. This is why the contents of dreams that occur during REM sleep are so vividly and accurately recalled.

THE SLEEP CYCLE

Under normal circumstances, an individual falling asleep passes into stage I of SWS first. There is then a natural progression from stage I through stage IV sleep (Fig. 18-2). After spending about 15 minutes in deep sleep, the depth of sleep decreases to stage II. REM sleep is entered from this stage of sleep. As indicated in Figure 18-2, the first cycle of SWS takes about 90 minutes and the period of REM sleep lasts for approximately 15 to 20 minutes.

This cycle repeats itself throughout the night. However, the periods of SWS gradually shorten, while the episodes of REM sleep lengthen during the night. Also, after the first or second cycle of SWS, little or no time is spent in the deeper stages of sleep. Brief periods of wakefulness occur periodically during the night, but these do not usually interfere with the progression of the sleep cycle.

Changes in Sleep Patterns During Development

Infants spend about 16 hours of every day sleeping. This time is reduced to about 10 hours by the age of 6 or 7 and to about 7 or 8 hours by adulthood. As one ages, the duration of sleep decreases, and during the sixth or seventh decade of

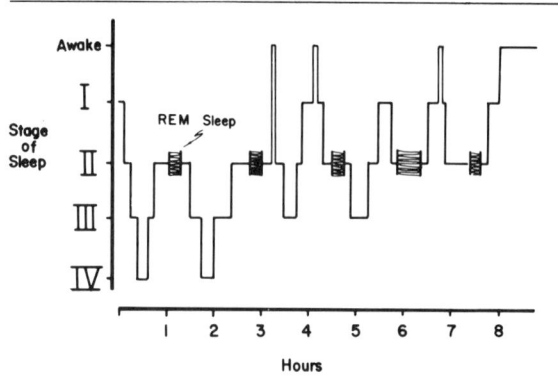

Fig. 18-2. During each sleep cycle the stages of slow wave sleep become progressively deeper and lighter. Occasional periods of wakefulness occur during a normal sleep cycle.

life individuals rarely spend more than 5 or 6 hours sleeping.

Interestingly, almost 70% of the total sleep time in infants is REM sleep, while in adults this drops to less than 25%. Thus, as one passes through childhood, almost all of the reduction in sleep time is due to a loss of REM sleep. Later, as an adult, the REM sleep time remains constant, and the loss in sleep is due to a reduction in the time spent in the deeper stages of SWS. The dominance of REM sleep during infancy and childhood has led to the theory that REM sleep may be important for the proper organization of brain circuits during development.

THE PHYSIOLOGY OF SLEEP

The neural mechanisms underlying the sleep-wake cycle are not fully understood. However, it is clear that there are at least three systems involved, one for producing wakefulness and one for each of the two states of sleep.

Anatomical Structures Involved in Sleep

Wakefulness, as indicated earlier, is maintained by the ascending reticular activating system (ARAS). The ARAS is a group of ascending pathways originating in the reticular formation and terminating within the thalamus. Those neurons within the thalamus receiving an input from the ARAS send diffuse projections to the entire cortex. Activity within this pathway is primarily responsible for producing the synaptic activity recorded by the EEG. In the absence of stimulation by the ARAS, the thalamus oscillates at a basic frequency of 10 Hz. This is the cause of the alpha rhythm. When activated by the ARAS, the neurons within the thalamus increase their firing rate, producing the beta rhythm of the EEG.

Stimulation of the ARAS results in an increase in the physiological state of arousal in addition to a change in the EEG. In this state, the brain is prepared to receive and interpret information. Since all of the sensory systems have projections to the ARAS, any one of them can initiate an arousal response. Damage to this system prevents an individual from becoming aware of his or her surroundings because sensory stimulation, although able to evoke activity in the specific sensory receiving areas in the cortex, cannot produce arousal. At one time it was thought that sleep was simply due to the loss of arousal. If external stimulation or internal thought processes were not sufficient to maintain activity in the ARAS, sleep followed. However, it is now clear that sleep is not just the withdrawal of excitation but is an active process resulting from the stimulation of brain stem pathways responsible for initiating the sleep state.

The anatomic location of the sleep centers has not been conclusively identified. Neurons in the locus ceruleus, located in the dorsal pons, have been implicated in producing at least some of the physiological characteristics associated with REM sleep. These neurons utilize norepinephrine as a transmitter and may project rostrally to stimulate rapid eye movements and caudally to inhibit skeletal muscle tone. It is not clear, however, whether it is these or the neighboring acetylcholine-containing neurons that are most responsible for REM sleep.

Similarly, the identification of the neurons responsible for producing SWS are not known for certain, although the pontine raphe nuclei are thought to be involved. These neurons, located in the upper pons, use serotonin as their neurotransmitter. When they are destroyed or the synthesis of their neurotransmitter is prevented by the administration of drugs, SWS is prevented.

DISORDERS OF SLEEP AND WAKEFULNESS

Although the physiological basis for sleep is not understood, it is quite clear that sleep is necessary to maintain normal behavior. Individuals who are deprived of sleep for long periods of time suffer from severe behavioral deficits. Reaction times are decreased, learning new information becomes difficult, and normal thought patterns are disturbed. The delusions and hallucinations that sometimes accompany sleep deprivation closely resemble acute schizophrenic episodes. Fortunately, all of the behavioral signs of sleep deprivation usually disappear when normal sleep patterns are restored.

There are separate requirements for slow wave and REM sleep. Thus, if an individual is selectively deprived of REM sleep, then the time spent in REM sleep during the next sleep cycle is increased above normal until the deficit in REM sleep is corrected. Similarly, if stage IV SWS is prevented, the individual spends an increased period of time in deep sleep when allowed to return to sleep. Also, the characteristics of sleep disorders that affect REM sleep are different from those that affect SWS.

Insomnia

The most common sleep disorder is *insomnia*, the inability to obtain an adequate amount of sleep. This is difficult to define in terms of the amount of time spent sleeping, since normal sleep periods vary from as little as 3 hours to as many as 9 or 10 hours per night. Additionally, many individuals complaining of an inability to sleep are often found to sleep quite normally during sleep laboratory studies. Thus, the diagnosis of insomnia is usually not made until the signs of sleep deprivation are present. Insomnia is usually secondary to some other disorder, such as discomfort, pain, or anxiety, and disappears when these problems are eliminated.

Prescribing drugs for the treatment of insomnia is often not helpful, as the sedative effects of sleeping pills remain with the patient throughout the day and thus interfere with normal waking activities. In addition, these drugs cause a reduction in the amount of time spent in REM sleep and can cause symptoms of REM sleep deprivation, such as agitation and hallucinations, to occur.

Disorders of SWS

Nocturnal enuresis (bed-wetting) and somnambulism (sleepwalking) occur during SWS and not REM sleep. The dreams that patients report in association with these sleep disorders only occur if the individual is allowed to remain asleep until the next REM sleep period occurs.

Night terrors are also associated with slow wave and not with REM sleep. These episodes, more common in children than adults, are characterized by the individual becoming suddenly awake, covered with perspiration, and staring wide-eyed into space or screaming, as if terrorized. Although it is natural to assume that these episodes are caused by bad dreams, sleep laboratory studies have shown that they do not occur during REM sleep and that upon waking, there is no recall of the attack nor of any dream preceding it. True nightmares, or frightening dreams, occur during REM sleep. Often these episodes include memories of events that took place during the waking state. Traumatic or highly stressful experiences can initiate nightmares. If they occur on a regular basis, psychiatric intervention may be required to eliminate them.

Disorders of REM Sleep

The most common disorder associated with REM sleep is *narcolepsy*. This disease is characterized by irresistable urges to sleep that occur during the day. The sleep periods only last for ten to fifteen minutes and are always of the REM type. A disorder closely associated with narcolepsy is *cataplexy*. During an attack of cataplexy, the person suddenly becomes paralyzed and falls to the ground. Although unable to move, there is no loss of consciousness and the individual is aware of what is happening. These episodes generally last for no more than a minute.

The dreams associated with the REM sleep periods of narcolepsy often take on an hallucinatory nature, because the individual passes suddenly from the waking state to the sleep state and returns to wakefulness totally alert. Sometimes the dreams appear to be occurring even while the individual is awake from all other behavioral signs. The paralysis of cataplexy is presumably caused by the same mechanisms that produce the profound hypotonia normally associated with REM sleep. The separation of dreams and paralysis from the unconsciousness of sleep indicate that there are probably separate neurophysiologic mechanisms for producing each of them. Although the cause of narcolepsy is not known, amphetamine and amphetamine-like drugs have been found effective in preventing or reducing symptoms.

19 Cognitive Functions of the Brain

The previous chapters of this text were concerned with the mechanistic functions of transmitting sensory information into the brain and the execution of commands by the motor control system. However, the most remarkable aspects of human neurophysiology are the processes by which we become aware of our sensory environment, consciously monitor our behavior, and communicate our thoughts and desires to others. These cognitive functions are performed by the association areas of the cerebral cortex and are the subject of this chapter.

Human brains are distinguished from brains that evolved earlier in phylogenetic history not only by their much larger size, but also by the proportion of brain tissue that is not devoted to specific sensory or motor tasks. For example, almost all of the cerebral cortex in rodent brains is devoted to the reception of sensory information or the issuing of motor commands. In human brains, only a small portion of the cortex is devoted to these tasks. All other areas of the brain can be considered as association cortex.

THE FRONTAL LOBE

The association areas of the frontal lobe, along with the limbic system, are primarily associated with our emotional and motivational state. The first recorded lesion of the frontal lobe was suffered by a mining engineer named Phineas Gage, who was struck by an iron bar during a dynamite explosion in 1848. The bar entered his skull and caused massive damage to his frontal lobe. The results of this lesion were surprising at the time, because it did not cause a major loss in mental ability, as was then expected. Instead, the lesion caused a profound change in his personality. He became unable to control his emotions and behaved in an ill-mannered, erratic, self-indulgent manner.

These characteristics were similar to those reported to have occurred in others with frontal lobe damage caused either by injury or as a result of the removal of a frontal lobe brain tumor. These syndromes of frontal lobe damage provided the rationale for the introduction of surgical frontal lobotomies for the control of severe depressive or psychotic behavior in 1935. Thousands of these operations were performed before the resulting loss of personality was judged to be too great a price to pay for the relief of the psychological disorder.

Frontal lobotomies (in which actually only the connections between the prefrontal lobe and the rest of the brain are severed) are still performed for the treatment of intractable pain in terminally ill patients. The sensation of pain does not disappear after the surgical procedure but the emotional suffering that accompanies the pain does. There is also a characteristic loss of anxiety and concern about the future that is quite palliative.

The emotional deficits caused by frontal lobe damage may be related to its disconnection from the limbic system and hypothalamus. The limbic system is an ill-defined group of brain centers and pathways that connect the prefrontal lobe to parts of the hippocampus, amygdala, and hypothalamus. The limbic system is thought to be concerned with the expression of emotional behavior.

The hypothalamus, occupying less than 1% of the brain mass, is an enormously complex area, controlling such diverse activities as eating and drinking, sexual behavior and reproduction, temperature regulation, coordination of the endocrine and autonomic nervous systems, learning, and emotion. It is beyond our scope to describe the details of the behaviors that are controlled by the hypothalamus and other limbic system centers. However, we should note that these functions are probably under inhibitory control by the frontal lobe and are released from inhibition when the frontal lobe is damaged or disconnected from the limbic system, producing the behaviors characteristic of the frontal lobe syndrome previously described.

THE TEMPORAL LOBE

The association areas of the temporal lobe are primarily related to memory and language. Memory functions are performed with the aid of the hippocampus, an area of the archicortex (having only three layers of cells rather than the six layers that characterize the neocortex) buried within the temporal lobe. Electrical stimulation of the hippocampus during surgery for the relief of temporal lobe epilepsy has produced some startling effects. For example, patients have reported recalling past events, not as an abstract memory, but as if they were occuring all over again.

Memory deficits occur as a result of damage to the hippocampus. These usually do not interfere with the recall of past events but with the storage of new information. The process of remembering probably involves three steps, an immediate type of memory that lasts only a few seconds (for example, knowing at the end of a sentence what was read at the beginning), a short term memory that lasts for about 20 minutes and is used to remember information required for current activities, and long-term memory that lasts essentially forever and is used for archival storage.

The hippocampus is involved in maintaining information in short-term storage until it can be permanently filed in a long-term memory storage site. Korsakoff's syndrome, observed in alcoholics, is a loss of the ability to transmit information from short-term to long-term storage sites that results from damage to the temporal lobe and hippocampus. Generally, there is no loss of long-term memory. Alzheimer's disease, the development of premature senility characterized by an inability to form new long-term memories, also appears to be caused by damage to the hippocampus.

The mechanism by which memory is held in short-term or long-term memory is not at all understood. Most theories assume that short-term memory is kept as some sort of reverberating neuronal circuit within the brain, because it is particularly susceptible to loss by electroconvulsive shock. Long-term memory, which seems to remain as a permanent part of cerebral function, is thought to be stored as an RNA-synthesized protein or as a growth of synaptic connections among neurons responsible for producing memories. At the moment, however, there is no real well-developed concept of how or even where memory is stored within the brain.

Language functions of the brain are better understood than most other cognitive functions because of the discrete deficits in language ability associated with specific brain lesions. These deficits in language ability are called *aphasias*. They are limited to the loss of language ability without any other loss of sensory, motor, or mental function. Two different types of aphasias have been identified; both are named for their discoverers, Broca in 1861 and Wernicke in 1874. Both types are associated primarily with lesions of the left side of the brain. About 90% of the population has language function lateralized to the left side of the brain. The different functions of the left and right hemispheres of the brain are discussed later in this chapter.

Broca's aphasia results from damage to the posterior frontal lobe just beneath the mouth regions of the motor cortex (see Fig. 17-1), and is characterized by the inability to perform the motor tasks associated with speech. There is no deficit in the motor control system for articulation, which is called *dysarthria*. However, the patient suffering from Broca's aphasia cannot generate the neural code for commanding the motor control system to produce language. As there is no defect in understanding language or in mentation, this is particularly frustrating for patients, since they know what they wish to say but are unable to express it. Speech is halting, not fluent, and lacks its normal cadence. Broca's aphasia is also called an expressive aphasia because of the inability to produce spoken language.

Wernicke's aphasia results from damage to the temporal lobe just behind the auditory areas (see Fig. 17-1). This type of aphasia is associated with the inability to understand language. There is no difficulty with generating the motor commands for speech. As a result, the aphasic is able to speak quite naturally but because there is no understanding of language, the speech lacks any real meaning. This sort of aphasia is also called a receptive aphasia because it is the understanding of language that is lost.

If the arcuate fasciculus connecting Wernicke's area to Broca's area is damaged, understanding of language is intact. However, the ability to transfer that knowledge from Wernicke's area to Broca's area is lost. As a result, speech has the normal rhythm but meaninglessness associated with Wernicke's aphasia.

Either because these two brain areas do not have exclusive control of language or because of the diffuse nature of most brain damage, aphasias are not purely expressive or receptive. However, these areas do seem to be predominantly concerned with language functions. Failure of one type of sensory information to reach these areas results in a loss of ability to use that modality of sensory communication for the comprehension of language.

For example, a lesion within the left occipital cortex that also involves the corpus callosum prevents visual information from reaching the language centers. As a result, written language becomes incomprehensible. The individual with such a lesion is able to see and even copy words from a book. However, the ability to understand the written words is lost. An analogous loss of the ability to understand speech occurs when damage to the left speech areas prevents sounds from reaching the language center.

THE PARIETAL LOBE

The association areas within the parietal lobe are involved in the generation of the neuronal activity required for issuing commands to the motor control system. Damage to this area results in a clinical syndrome called *apraxia*, a loss of the ability. It is in many ways analogous to the defects seen in Broca's aphasia. However, there are ability. It is in many ways analogous to the defects seen in Broca's aphasia. However, there are some differences. The most obvious of these is that in apraxic patients the motor tasks that cannot be performed voluntarily can be performed quite normally if elicited as part of an automatic behavior. For example, if asked to tie his or her shoes, the apraxic patient is unable to do so, even though he or she knows what is expected and can explain how it is done. However, if this same patient is putting on his or her shoes, he or she has no trouble tying the laces.

THE SPLIT BRAIN

One of the most obvious features of the cerebrum is that it is divided into two hemispheres that are connected by the corpus callosum. It had been known since the middle 1800s that language ability was localized on the left hemisphere in the overwhelming majority of persons. However, it was not until the 1960s that an appreciation for the intellectual functions of the right brain was developed. This occurred as a result of testing patients who had the corpus callosum surgically cut as a treatment for intractable epileptic seizures. Originally, no intellectual or emotional changes appeared to result from the operation, and it was assumed that communication between the two brain hemispheres was not required for normal behavior. However, further testing revealed that the right hemisphere has its own important intellectual functions and that communication between the left and right hemispheres occurs even in the absence of the corpus callosum.

In order to test the intellectual capacities of the two hemispheres, it was necessary to supply separate information to them. This was accomplished using the experimental arrangement illustrated in Figure 19-1. A picture presented on the left side of the screen is projected only to the right hemisphere, while a picture presented to the right side of the screen travels to the left

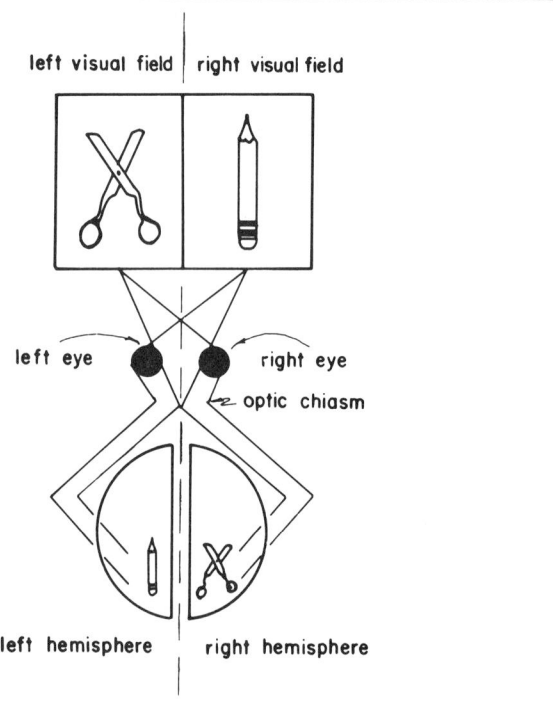

Fig. 19-1. The apparatus used to test the perceptual capabilities of the left and right hemispheres. Objects within the left visual field project to the right hemisphere, while those in the right visual field project to the left hemisphere.

hemisphere. If the image of a pencil is projected to the right visual field so that it projects to the left occipital cortex, the person has no trouble naming it. However, if the image is placed in the left visual field, the person either denies that any image appeared or else randomly selects a name to describe it. However, if the person is asked to use his or her left hand (controlled by the right hemisphere) to pick out the object corresponding to the image projected to the right hemisphere, the scissors are immediately retrieved. Thus, although the person is unable to communicate about the image projected into the right hemisphere, the visual information can nonetheless be used to guide behavior.

The role of the right hemisphere is difficult to

ascertain because it cannot generate the motor commands required for spoken language. However, various experiments have been conducted to indicate that the main task of the right hemisphere is to decode spatial images, particularly when these involve sensorimotor coordination. The right hemisphere is much more successful then the left in identifying an object placed in the hand or in making a drawing of a geometrically complex object.

One of the complications in separating the functions of the two hemispheres is the ability of the left hemisphere to provide a rational explanation for behaviors initiated by the right hemisphere. For example, when a word such as "rub" was projected to the right hemisphere of an individual who had some rudimentary right-sided language ability, the individual rubbed one hand with the other. However, when asked what the command was, the patient responded by saying "itch". In other words, the left hemisphere, having no idea why the right hemisphere directed one hand to rub the other, had no difficulty in fabricating a reason for it.

A more striking example resulted from an experiment in which the images projected to the two hemispheres of a person without a corpus collusum were different. The left hemisphere saw a claw of a chicken, while the right hemisphere saw a snow-filled scene. When the person was asked to choose a picture most closely related to what he had observed, the left hemisphere directed the right arm to a picture of a chicken, while the right hemisphere caused the left arm to choose a shovel. He was then asked to explain the reason for his choices. The left hemisphere, having no idea why the right brain chose the shovel, had no trouble making up a reason. Without any hesitation he explained that he saw a claw, so he chose the chicken, and everyone knows that a shovel is used to clean out the chicken shed.

The true cognitive capabilities of the right hemisphere will not be known until better testing methods are developed. However, until then we are free to explain away otherwise incomprehensible behaviors by saying "the right brain made me do it".

Glossary

Accommodation When an image is brought close to the eye, it is kept in focus by increasing the refractive power of the lens. This increase is accompanied by convergence of the eyes and constriction of the pupils. Together these responses are called *accommodation for near vision*.

Activation An increase in membrane conductance due to the opening of gating molecules during depolarization.

Active state In muscle, the period of time during which intracellular calcium concentration is high enough to permit cross-bridge cycling to occur.

Adaptation The decrease in frequency of firing in a sensory neuron despite continued presence of the stimulus.

Adenylate cyclase A membrane-bound enzyme that synthesizes cyclic-AMP from ATP when activated by a stimulus (usually a hormone or neurotransmitter).

Adequate stimulus The type of stimulus for which a receptor is designed to respond. Other receptors also may respond to this stimulus but require much higher energy levels to do so.

All-or-none response If threshold is achieved, the response to a stimulus is always the same.

Angstrom A unit of length equal to 10^{-10} meters or 0.1 millimicrons (mμ).

Ataxia Pathologic loss of motor coordination usually caused by lesions to the cerebellum.

Autogenic inhibition The disynaptic reflex initiated by the Golgi tendon organ (GTO). When the GTO is stretched, it causes extrafusal muscle fibers to contract.

Capacitive current The portion of the current applied to a membrane that serves to charge the membrane capacitance. The rate at which the membrane potential changes is proportional to how quickly the membrane capacitance is charged.

Chemically excitable channels Channels that open or close in response to a chemical substance binding to a receptor. Unlike the electrically excitable channels, these channels cannot

Chemically excitable channels—*continued* give rise to the regenerative response necessary for the production of an action potential.

Coactivation Simultaneous activation of alpha and gamma motoneurons prevents unloading of muscle spindles during a muscle contraction.

Corollary discharge When a motor task is issued to the motoneuron pools of the spinal cord and brain stem, corollary discharge is directed to the other components of the motor control system. The information obtained is used to aid in the execution of the movement.

Cross bridge The globular end of the myosin molecule that attaches to the actin molecule during contraction and by bending causes the thin filament to slide across the thick filament.

Decerebrate rigidity A condition resembling spasticity that occurs when the portions of the brain responsible for inhibiting activity of the brain stem reticular formation are damaged.

Denervation supersensitivity An increase in the excitability of a muscle or neuron following the loss of its normal innervation. Generally caused by an increase in the number of synaptic receptor sites.

Depolarization Making the membrane more positive. Because the absolute magnitude of the membrane potential is reduced, the membrane potential is considered to be reduced by depolarization.

Diffusion The random oscillation of particles in solution from a region where they are more concentrated to a region where they are less concentrated.

Electrically excitable membrane A membrane capable of producing an action potential.

Electrochemical potential The energy driving ions across a membrane, consisting of the energy contained in the concentration gradient for the ion (the chemical potential) and the energy contributed by the membrane potential (the electrical potential).

Electroneutrality The state at which total number of negative charges on ions in solution equals the total number of positive charges. Thus, when an ion crosses a membrane its charge must be balanced. Energy, in the form of a membrane potential (voltage), is acquired when negative and positive charges are separated from each other by a membrane.

End plate The specialized region of a skeletal muscle fiber where neuromuscular transmission occurs.

End plate potential The depolarization of a skeletal muscle fiber during synaptic transmission.

Equilibrium potential The number of millivolts required to prevent the net flux of an ion across a cell membrane by diffusion. It also represents, in electrical units (millivolts), the energy contained in a concentration gradient. It is calculated using the Nernst equation.

Excitatory postsynaptic potential (EPSP) A membrane depolarization produced by the action of a neurotransmitter. The end plate potential is one example of an EPSP.

Exocytosis A process in which vesicularly bound material is released from a cell. During exocytosis the vesicle coalesces with the cell membrane, allowing its contents to escape into the extracellular fluid.

Extrafusal muscle fibers The individual fibers of a muscle that are used to generate force and move a limb. Extrafusal muscle fibers are distinguished from intrafusal muscle fibers contained within the muscle spindles.

Extraparametal system The portion of the motor control system that originates within the cortex but does not descend to the spinal cord in the corticospinal (parametal) trap.

Feature detectors Neurons within the brain that are involved in sensory coding. When activated they give rise to a complex perceptual quality.

Focal length Parallel rays of light entering a lens converge at a point behind the lens called its *focal point*. The distance between the focal point and lens is called the *focal length*.

Gamma bias The excitability of the gamma loop, determining the sensitivity of the stretch reflex to changes in muscle length.

Gamma loop The reflex pathway through which sensitivity of the intrafusal muscle fibers can be controlled.

Gap junction A junction between two cells through which electrical current and small ions can pass.

Generator potential The change in membrane potential of a receptor in response to a stimulus. The term is often used interchangeably with *receptor potential*.

Golgi tendon organ The receptor found within the tendonous insertions of muscles. It detects changes in tension produced by the muscle.

Graded response The magnitude of the response proportional to the strength of the stimulus; for example, the number of quanta released during synaptic transmission or the amount of light shining into the eye.

h gate One of the two gates covering the sodium channels. The h gates close during membrane depolarization, causing sodium conductance to decrease. The h gates are also referred to as *inactivation gates*.

Hyperpolarization Making the membrane potential more negative than its resting potential. Because the absolute magnitude of the membrane has increased, the membrane potential is considered to be increased by hyperpolarization.

Inactivation Reduction in membrane conductance due to the closing of a gating molecule on a channel during membrane depolarization. Closing of the h gates during depolarization inactivates the sodium channels.

Inhibitory postsynaptic response (IPSR) A decrease in membrane excitability caused by a synaptic transmitter. It is usually, but not always, caused by hyperpolarization of the membrane.

Intrafusal muscle fiber A modified muscle fiber contained within the muscle spindle that detects changes in muscle length and participates in the reflex control of muscle contraction.

Labeled line A neuronal pathway involved in sensory coding. No matter how the pathway is activated, whether through stimulation of its receptor by an adequate stimulus or stimulation of the brain by an electric current, the same sensation is perceived.

Law of mass action Used in the field of enzyme kinetics to describe a situation in which the rate of a reaction is proportional to the concentration of substrate, product, and enzyme. When applied to the carrier-mediated transport of particles, it implies that the rate of transport increases rapidly as the concentration of the particle increases and then levels off, reaching a maximum when the concentration of the particle exceeds the concentration of the carrier.

m gate One of the two gates covering the sodium channel. The m gate opens during membrane depolarization, causing sodium conductance to increase. It is also referred to as an *activation gate*.

Motor unit The functional unit used by the motor control system to carry out muscular activity, consisting of an alpha motoneuron and all of the muscle fibers it innervates.

Muscle twitch Rise in muscle force due to the events following a single action potential. The muscle twitch begins when an action potential causes calcium to enter the myoplasm and concludes when the calcium is resequestered by the sarcoplasmic reticulum.

Myelinated fiber A sensory or motor axon covered with an insulating sheath of Schwann cell membranes. The nerve membrane is exposed to the extracellular fluid at the nodes of Ranvier, which are located between the individual Schwann cells covering the axon.

Myofibril The longitudinal unit of a striated muscle. Individual myofibrils (which are several microns in diameter) are separated from each other by the surrounding sarcoplasmic reticulum.

n gate The gate covering the potassium channel. The n gate opens during membrane depolarization, causing potassium conductance to increase. It is also referred to as an *activation gate*.

Negative feedback A condition in which the response to a stimulus is reduced through the action of the response itself as an inhibitory stimulus.

Nernst equation An equation used to calculate the equilibrium potential for an ion.

Nodes of Ranvier The spaces between individual Schwann cells in a myelinated neuron at which the axonal membrane is exposed to the extracellular fluid. Action potentials propagated along the myelinated axon occur only at these nodes.

Nystagmus The reflex movement of the eyes in a direction opposite to the movement of the head, occurring to maintain fixation on an object in the visual field. Spontaneous nystagmus or its absence is usually a sign of vestibular or cerebellar disease.

Parallel elastic component (PEC) The component of a muscle fiber that offers resistance when a muscle is stretched. In striated muscle the cross bridges do not contribute to the PEC.

Parametal system The portion of the motor control system that originates within the cortex and reaches the spinal cord via the corticospinal (parametal) trap.

Phasic receptor A receptor that rapidly decreases its rate of firing despite continued presence of the stimulus.

Phosphorylation The addition of inorganic phosphate to a receptor or channel. Phosphorylation is an important mechanism by which intracellular proteins are regulated. It is catalyzed by a kinase enzyme.

Quantal release During synaptic transmission each vesicle releases a similar amount of transmitter, called a *quantum*.

Reciprocal innervation The circuitry of the nervous system is organized so that when a motor task is executed, the excitation of the antagonist and inhibition of the antagonist occur together. Because of this reciprocal innervation, a movement can occur without interference from the antagonist.

Receptive field The region of a sensory organ, such as the skin or eye, which influences the firing of a single sensory neuron when stimulated.

Reflex Stereotypic motor behavior produced by the nervous system in response to a stimulus.

Refractive power The power of a lens to bend a ray of light. The greater the degree of curvature and the slower the speed of light in the lens, the greater its refractive power. The refractive power in diopters is equal to the reciprocal of the focal length in meters.

Refractory periods The periods following an action potential during which membrane excitability is decreased. During the absolute refractory period, an action potential cannot be elicited no matter how great the stimulus. During the relative refractory period, an action potential can be elicited but the stimulus must be greater than normal.

Regenerative (positive feedback) response A condition in which response to a stimulus is prolonged by becoming a stimulus to itself. The upstroke of the action potential is a regenerative response because the depolarization of the membrane (the response) causes (i.e., becomes the stimulus for) further depolarization.

Reversal potential The potential at which net current flow through a membrane is zero. It is most commonly used to describe the effect of a synaptic transmitter on membrane conductance. In this context, the reversal potential is the potential at which activation of the membrane by a synaptic transmitter would not cause a change in membrane potential.

Saltatory conduction The propagation of the action potential in a myelinated nerve in which the action potential jumps from one node of Ranvier to another.

Sarcomere The functional unit of striated muscle, consisting of the thick and thin filaments between two adjacent Z bands in a myofibril.

Sarcoplasmic reticulum (SR) A longitudinal network of tubules found in striated muscle that plays an important role in excitation-contraction coupling. The SR contains a high concentration of calcium which is released when an action potential is propagated along the T-tubules.

Series elastic component (SEC) The component of a muscle that must be stretched before the force developed by the cross bridges can be transmitted to the bones. The SEC is stretched during repetitive cross-bridge cycles.

Size principle During the performance of a motor task, small alpha motoneurons are recruited before large ones. This orderly arrangement of recruitment occurs because small alpha motoneurons are easier to excite than large ones.

Sodium-potassium pump Sodium-potassium ATPase, a membrane-bound enzyme that uses the energy derived from the hydrolysis of ATP to transport sodium out of and potassium into a cell against their concentration gradients.

Space constant (λ) A measure of how far a change in membrane potential can spread along the membrane. The larger the space constant, the more effectively a membrane can propagate an action potential. The space constant increases as cell diameter and membrane resistance increase.

Spasticity A pathologic increase in the tone of antigravity muscles caused by the loss of inhibition to the brain stem reticular formation.

Steady-state potential The membrane potential resulting from the constant flow of ions across the membrane. Energy is consumed in maintaining the concentration gradients required to prevent the flow of ions from changing.

Stretch reflex The monosynaptic reflex initiated by the intrafusal muscle fibers within the muscle spindle. It is elicited when the intrafusal muscle fiber is stretched and causes the extrafusal muscle fibers to contract.

Summation In synaptic transmission, the change in membrane potential produced by a neurotransmitter is increased by summation of the effects of two or more synaptic potentials. In muscle, the strength of a contraction is increased by prolonging the duration of the active state.

Synapse The site at which cells communicate with each other.

Synaptic vesicle A membrane-bound vesicle found in an axon terminal near the synapse. Each of the axon's many vesicles contains a similar amount of neurotransmitter, which is released by exocytosis during synaptic transmission.

T-tubules A series of tubules formed by the invagination of striated muscle fiber membranes. T-tubules serve to propagate action potentials into the interior of the muscle fiber and stimulate release of calcium from the sarcoplasmic reticulum.

Tetanus Summation of muscle contraction in which force rises smoothly to the maximum value the muscle is capable of generating.

Threshold (1) The membrane potential at which an action is elicited or (2) the magnitude of the stimulus voltage or current that must be applied to the membrane to cause an action potential to be elicited.

Time constant (τ) A measure of how rapidly a stimulus can cause the membrane potential to change. The larger the time constant, the more slowly the membrane potential changes. Its magnitude increases as membrane resistance and capacitance increase.

Tonic receptor A receptor that continues to discharge action potentials for as long as the stimulus is maintained.

Topographic representation Organization of the sensory or motor area of the nervous system in which contiguous receptor fields, muscle groups, or closely related sensory qualities are

Topographic representation—*continued* represented by adjoining regions of the brain or spinal cord.

Transducer membrane The component of a membrane responsible for detecting a stimulus and converting it into an electrical response called the *receptor potential*.

Transference The total membrane conductance is the sum of the conductances for each ion permeable to the membrane. The percentage of the total membrane conductance due to a single ion is the *transference* for that ion.

Unloading of muscle spindles The decrease in tension on the intrafusal fiber within the muscle spindle that occurs during a contraction. This unloading of the muscle spindle reduces its ability to detect changes in muscle length and to control movement.

Visual acuity A measure of how well the eye can identify objects. Visual acuity is usually measured by determining the smallest letter that patients can recognize on an eye chart placed 20 feet away.

Index

Index

A band, myofibril, 44–45
A fiber axon, 66
Accommodation
  in image formation, 92–93
  for near vision, 92
Acetyl Co A, 32
Acetylcholine
  heart rate and, 41
  inactivation of, 34
  in myasthenia gravis, 34–35
  in neuromuscular synapse, 31–34
Acetylcholinesterase, 34
Acetyltransferase, 32
Actin, 45–46
Action potential, 19–28
  active propagation of, 27
  in cardiac muscle, 52
  components, 20
    conduction velocity, effect of space and time on, 27–28
  downstroke, 19–22
    inhibition of, by end plate potential, 35
    initiation of, by excitatory postsynaptic potential, 37
  ionic gates, 22–23
  membrane capacitance, 27
  motoneuron, 130
  overshoot, 19, 20
  passive current spread, 26–27
  passive response and, 20
  propagation of, 26–28
    in unmyelinated axon, 28
  receptor, 60
  role of gates in, 22–26
  saltatory conductance, 28
  of skeletal muscle, 46–47
  space constant, 27
  undershoot, 19, 20
  upstroke, 19, 20–21
Activation gate, 22
Active transport
  direct, 5–6
  indirect, 6–7
  sodium-potassium pump in, 5–6
Acupuncture, 75
Adaptation
  dark, 101–102
  perceptual, 62–63
  receptor, 60–61

Adapting receptor, 60
Adenosine triphosphate. *See* ATP
Adenylate cyclase, cyclic AMP and, 4
Adiodochokinesia, 167
Adrenergic postganglionic response, 42
Afterdischarge reflex, 140
Ageusia, 79
Alerting response, sleep, 179
Alpha blocking, in sleep, 179
Alpha receptor, 42
Alpha rigidity, 149, 152
Alzheimer's disease, 184
Amblyopia, 104
Ametropia, 93
Amino acid, active transport, 6–7
Anomalous color vision, 104–105
Anosmia, 79, 84
Antigravity muscle, 148–149
Aphasia, 185
Apraxia, 185–186
Aqueous humor, 88–89
Archicerebellum, 163
Archicortex, 184
Arcuate fasciculus, 185
Articulation, 185
Ascending reticular activating system, in sleep, 179, 180–181
Association cortex, 169
Ataxia, 167
Athetosis, 160
ATP (Adenosine triphosphate)
  in cross-bridge cycle, 47–48
  in muscle fibers, 131–133
  phosphorylation, 6
Atropine, in muscarinic blockade, 42
Audiometry, 121–122
Auditory pathway, 118, 119
Autogenic inhibition, 141
Autonomic nervous system, chemical transmission in, 40–42
Axoaxonic synapse, 39–40
Axon
  cerebellum, 164–166
  equivalent circuit, 26–27
  motoneuron, 130
  sensory
    classification of, 66
    sizes, 66
Axon hillock, summation, 38–39

Basal cell, olfactory, 84
Basal ganglia, 157–161
   anatomical connections, 158–159
   function of, 157–158
   lesions of, 127, 159–160
   in motor control, 126
   motor deficits due to, 159–161
Basilar membrane, 112, 114–115, 117
Battery
   cell membrane, 9
   ionic, 13–14
Bed-wetting, 182
Behavior
   olfactory stimulation and, 83
   sleep deprivation and, 181
Beta receptor, 42
Betz cell, 172
Biochemical flux, 13–14
Biological rhythms, 177
Bipolar cell, retinal, 96
Black widow spider venom, 34
Blind spot, 87–88
Blindness
   color, 104–105
   night, 96
Bony labyrinth, 112, 113
Botulinus toxin, 34
Brain
   cognitive functions, 183–187
   electrical stimulation, in pain control, 75
   frontal lobe, 183–184
   hemispheres, 186–187
   left hemisphere, 186
   in pain sensation, 73
   parietal lobe, 185–186
   right hemisphere, 186–187
   split, 186–187
   in taste sensation, 80
   temporal lobe, 184–185
   in thermoregulation, 77
   in touch sensation, 68–69
Brain stem, 147–155
   in motor control, 126, 127
   reticular formation, 147–149
   tonic neck reflexes, 154–155
   vestibular apparatus, 149–151
   vestibular reflexes, 151–154
Broca's aphasia, 185
Brodmann's areas, 118

C fiber axon, 66
Calcium
   in cardiac muscle, 52

Calcium—*Continued*
   -induced calcium release hypothesis, 47
   intracellular, 7
   photoreceptor potential and, 96
   presynaptic terminal, 32
   release, calcium-induced, hypothesis, 47
   sarcoplasmic reticulum (SR) release, 46–47
   in smooth muscle contraction, 52–53
   in synaptic transmission, 31–32
   trigger, 47
Calcium channel, presynaptic membrane and, 31
Calmodulin, in smooth muscle, 53
Calsequestrin, 49
Canal of Schlemm, 89
Capacitive current, cell membrane, 12–13
Capacitor, cell membrane, 11
Cardiac muscle, activation, 51–52
Cataplexy, 182
Cataracts, 89
Caudate nucleus, 158
Cell membrane, 3–4
   action potential. *See* Action potential
   batteries, 9
   capacitance, 11
   channels, 10–11
   conductors, 9
   diffusion, 4–6
   electrical potential, 6, 7
   equilibrium potential, 11
   equivalent circuit, 9–11
   flux density crossing, 4–5
   ion selective channel, 4
   ionic channel, 10
   lipid bilayer proteins, 4
   Nernst ion potentials, 7
   Ohm's law and, 11–14
   potassium channel, 10
   potential, 11
   pumps, 11
   resistance, 10–11
   resting potential, 9–18
      calculation of, 15
      depolarization, 18
      hyperpolarization, 17–18
      ionic gradient effect on, 16–18
      recording the, 16–17
      sodium-potassium pump and, 15
      transference equation and, 15
Schwann cell, 28
selectivity filter, 10
sodium channel, 10

Cell membrane—*Continued*
  translocation across, 4–7
Central nervous system, chemical
      transmission within, 35–40
  action potential inhibition, 37
  at motoneuron, 35–36
  excitatory postsynaptic potential, 36–37
  inhibition, 38–39
  inhibitory postsynaptic potential, 37
  summation, 38
Cerebellum, 163–167
  anatomy of, 163–166
  lesions of, 166–167
  in motor control, 127–128
Cerebral cortex, 169–173
  association, 169
  cell columns, 172–173
  cell layers, 171–172
  motor areas, 169–171
    primary, 169–170
    secondary, 169–170
    supplementary, 169–170
  in motor control, 128
  movement control by, 171–173
Cerebrocerebellum, lesions to, 167
Cerebrum, hemispheres of, 186–187
Cerumen, 120
Chemoreceptor, 59
Chloralabe, 104
Chloride equilibrium, postsynaptic
    potential, 37
Cholecystokinin, 37
Cholinergic postganglionic response, 41–42
Chromophore, 59
Circadian rhythm, 177
Circumvallate papilla, 80
Clathrin, synaptic vesicle, 32
Climbing fiber, of cerebellum, 165
Cochlea, anatomy of, 112–113
Cochlear duct, 112, 113
Cochlear microphonic potential, 116–117
Cold
  paradoxical, 77
  sensation of, 76–77
Cold fiber, 76, 77
Colliculus, superior, 99
Color, perception of, 104
Color blindness, 104–105
Concentration gradient
  diffusion and, 4
  facilitated diffusion and, 5
  Nernst equation and, 7
Conductance, 9

Conductance—*Continued*
  action potential and, 19–22
Conduction, saltatory, 28
Conductor, cell membrane, 9
Cone, 94–95
  in dark adaptation, 101–102
Contraction, muscular. *See* Excitation-
    contraction coupling
Contrast, perception of, 102–103
Cornea, 88
  stroma, 88
Corollary discharge, cerebral cortex, 173
Corpus callosum, 186
Cortical column, visual cortex, 100
Corticobulbar tract, 128
Corticospinal tract, 128, 171
Cortilymph, 114
Crista, 149
Cross bridge, skeletal muscle, 45–46
Cross-bridge cycle, 47–48
  in cardiac muscle, 52
  in smooth muscle, 53
Crossed extension reflex, 140
Cupula, 150, 152
Curare, 42
Current
  capacitive, 12–13
  ionic, 13
  net pump, 15–16
Cutaneous sensation, 65–77
  Meissner corpuscles and, 66–67
  Merkel's disks and, 67
  Pacinian corpuscles and, 66
  pain receptors and, 71–75
  Ruffini nerve endings and, 66
  thermal receptors and, 75–77
  touch receptors and, 65–71
Cyanolabe, 104
Cyclic AMP, synthesis of, 4
Cytoplasmic channel, gap junction, 29–30

Dark adaptation, 101–102
Decerebrate rigidity, 148–149
Decibel, 110
Deiter's nucleus, 152
Delusions, sleep deprivation and, 181
Denervation supersensitivity, 34
Dense bar, presynaptic membrane, 31
Depolarization
  absolute refractory period in, 25
  all-or-none response in, 24
  local response, 24–25
  of membrane potential, 18

Depolarization—*Continued*
  passive, 20
  refractory period, 25–26
  relative refractory period in, 25, 26
  T tubule, 46–47
Depth perception, 105–106
Dermatome, 69
Desensitization, 34
Deuteranopia, 104
Dichromacy, 104
Diffusion, 4
  facilitated, 5, 6
Diplopia, 104
Disk, retinal, 94–95
Diurnal rhythm, 177
Dopaminergic pathway, in Parkinson's disease, 160
Dream sleep, 179–180
Driving force, cell membrane, 11
Dynamic receptor, 60
Dysarthria, 185
Dysgeusia, 79
Dysosmia, 79
Dystonia, 160

Ear
  inner, 112–115
  middle, 110–112
  wax build-up in, 120
Eardrum, 111
Elasticity, of skeletal muscle, 48–50
Electrochemical gradient, 13
Electroencephalogram
  alpha wave, 179
  beta wave, 179
  delta wave, 179
  in sleep monitoring, 178
  theta wave, 179
Electrogenic pump, cell membrane, 11
Electroneutrality, cell membrane, 12
Emmetropia, 93
Emotion
  frontal lobe and, 183
  frontal lobotomy and, 184
  olfactory stimulation and, 83
End plate
  neuromuscular, 31
  potential, 32–33
    in action potential inhibition, 35
    miniature, 33–34
Endolymph, 112
Endomysium, 43
Endorphins, 75
Enkephalins, 75

Enuresis, nocturnal, 182
Epilepsy
  electroencephalography in, 179
  temporal lobe and, 184
Equilibrium, establishment of, 12
Equilibrium state, cell membrane, 11
Equivalent circuit
  action potential, 21
  axon, 26–27
  cell membrane, 9–11
Erythrolabe, 104
Excitation-contraction coupling, 43–53
  cardiac muscle, 51–52
  skeletal muscle, 43–51
    activation of, 45–50
    length-tension relationship in, 48–49
    muscle relaxation in, 49–50
    parallel elastic component (PEC) in, 50
    relaxation in, 49–50
    series elastic component (SEC) in, 50
    sliding filament hypothesis in, 48
    summation in, 50–51
    tetanus and, 50–51
  smooth muscle, 52–53
Excitation-secretion coupling, in synaptic transmission, 31–32
Excitatory postsynaptic potential, 36–37, 39
  action potential initiation, 37
Extensor inhibitory area, brain stem, 148
Exteroceptor, 57
Extracellular fluid, ionic concentration of, 3, 7
Extrapyramidal system, 128, 171
Eye, 88–89
  anatomy of, 88–89
  far point of, 93
  image formation by, 91–92
  near point of, 93
  strain, 93
Eye chart, 101, 106–107
Eyeball, development of, 94
Eyeglasses, corrective, 93–94

Farsightedness, 93–94
Fast fatigable (FF) muscle, 131
Fatigue resistant (FR) muscle, 131
Feature detector, 63, 105
Fick's law of diffusion, 4–5
Flexor reflex efferent, 140
Floater, vitreous humor, 89
Flocculonodular lobe, 163
  lesions to, 166
Fovea, 99
  in depth perception, 105–106

Fovea—*Continued*
   feature detector, 105
   visual acuity and, 101
Frontal lobe, function of, 183–184
Fungiform papilla, 80–81

Gamma bias, 142–144
Gamma dynamic fiber, 137, 138
Gamma efferent fiber, 137, 138
Gamma loop, 142
Gamma motoneuron, 126–127
Gamma rigidity, 149, 152
Gamma static fiber, 137, 138
Gamma-aminobutyric acid (GABA), 37, 39
Ganglion
   autonomic, 40–41
   basal, 157–161
Ganglion cell
   in color blindness, 104–105
   in contrast perception, 102–103
   retinal, 96, 98–99
Gap junction, electrical transmission and, 29–30
Gate, action potential, 22–26
   activation, 22
   H, 22, 25
   inactivation, 22
   ionic, 22
   M, 22
   N, 22
   positive feedback response, 23
   regenerative response, 23
   sodium channel activation and, 23
   sodium channel inactivation and, 23–24
   time effect on, 23
Gate-control theory, of pain sensation, 75
Generator potential, 58
Geniculate nucleus, 97, 98
Glaucoma, 89, 107
Glial cells, 94
Globus pallidus, 158–159
Glomerulus, olfactory, 83
Glucose
   diffusion and, 4
   facilitated diffusion and, 5
Golgi tendon organ, 136, 144–145
Granular cell layer, of cerebellum, 165
Guanosine cyclic monophosphate, in photoreception, 95–96
Gustatory stimulus, 81

H band, myofibril, 44–45
H gate, 22
Hair cell, 113–117

Hair cell—*Continued*
   cochlear microphonic potential, 116–117
   receptor potential, 115–116
   transduction process, 116
   vestibular, 149
Hallucinations
   narcolepsy and, 182
   sleep deprivation and, 181
Hearing, 109–122
   clinical aspects of, 120–122
   hair cells in, 115–117
   inner ear in, 111–115
   loss, 120
      tests in, 121–122
   middle ear in, 110–112
   neural encoding in, 117–120
   sound stimulus, 109–110
Hearing aid, 121
Heart, acetylcholine effect on, 41–42
Heat, sensation of, 76–77
Heavy meromyosin, 45–46
Helicotrema, 112, 113
Hemiballismus, 159, 160
Hensen's cell, 113, 114
High gamma bias, 143–144
Hippocampus, function of, 184–185
Homeostasis, cellular, 3–7
Huntington's chorea, 160
5-Hydroxytryptamine, in pain inhibition, 75
Hyperopia, 93–94
Hyperpolarization, of membrane potential, 17–18
Hypokinesia, 159
Hypothalamus
   function of, 184
   in smell sensation, 83
   in thermoreception, 76–77

I band, myofibril, 44–45
Ia sensory fiber, 137
Ib afferent fiber, 136
Image formation, 89–94
   accommodation in, 92
   depth perception and, 105–106
   by eye, 91–92
   focal distance, 90
   focal plane, 90
   focal point, 90–91
   image distance, 90–91
   image plane, 90–91
   refractory power in, 90–91
Impedance mismatch, in hearing, 111
Inactivation gate, 22
Incus, 111

Inhibitory postsynaptic potential, 37, 39
Innervation, reciprocal, 141
Insomnia, 181–182
Insulin, 37
Intention tremor, 167
Interoceptor, 57–58
Intracellular fluid, ionic concentration of, 3, 7
Intraocular pressure, 89
    measurement of, 107
Ionic battery, charging, 13–14
Ionic channel, 10
Ionic current, 13
Ions, Nernst potentials of, 7
Irradiation reflex, 140
Isopotentiality, 16

Junctional pillar, 44–45

Kinocilium, 149–150
Korsakoff's syndrome, 184

Labyrinth, vestibular, 149
Language, ability deficits, 185
Lens
    anatomy of, 88–89
    refractory power of, 90–91
Light meromyosin, 45–46
Light ray
    in image formation, 90
    principal axis, 90
Limbic system
    function of, 184
    in smell sensation, 83
Lobotomy, frontal, 184
Local sign reflex, 140
Low gamma bias, 143–144

M band, myofibril, 44–45
M gate, 22
Macula, 150–151
Malleus, 111
Medulla
    in pain sensation, 73
    in touch sensation, 68–69
Meissner corpuscle
    in stimulus encoding, 68
    structure of, 66–67
Membrane capacitance, action potential, 27
Membranous labyrinth, 112, 113
Memory, temporal lobe and, 184
Merkel's disk, 65
    in stimulus encoding, 68

Merkel's disk—*Continued*
    structure of, 67
Meromyosin, 45–46
Mesopic vision, 102
Microphonic potential, cochlear, 116–117
Midget cell, foveal, 101
Miniature end plate potential, 33–34
Mossy fiber, of cerebellum, 165
Motoneuron
    alpha, 126, 129–130
    gamma, 126–127
    pool, 127
    size principle, 126
    synaptic transmission at, 35–36
Motor control system
    components, 126
    overview, 125–128
Motor deficit, basal ganglia lesions and, 159–161
Motor unit, 126
    muscle fiber types, 131–133
    physiologic control, 133–134
Müller cells, 94
Muscarinic receptor, 42
Muscle. *See also specific type*
    antigravity, 148–149
    cog wheel rigidity, 160
    lead pipe rigidity, 160
    reflexes, 140–141
    stretch, 138–139
    tremors, 161
Muscle fiber
    ATP in, 131–133
    classification, 131–133
    extrafusal, 136
    fast fatigable, 131
    fatigue resistant, 131
    innervation, 132–133
    intrafusal, 136–137
    red, 131
    white, 131
Muscle receptor, 135–139
Muscle spindle, 136, 137
    afferents, 137
    unloading, 138–139
Myelinated axon, propagation in, 28
Myasthenia gravis, 34–35
Myofibril, 44–45
Myofilament, 44–45
Myopia, 92–93
    floaters and, 89
Myosin, 45–46
    in smooth muscle, 53

Myosin light chain kinase, 53

N gate, 22
Narcolepsy, 182
Near vision, accommodation for, 92
Near-sightedness, 92–93
    floaters and, 89
Neck reflex, 154–155
Neocerebellum, 163–164
Nernst equation, 7, 11
    cell membrane battery and, 9
    derivation of, 12
Nervous system, pain originating within, 74–75
Neural pathway
    auditory, 118, 119
    in touch sensation, 68–69
Neuromuscular junction, 30–31
Neuron
    cellular homeostasis, 3–7
    equivalent circuit, 9–11
    as feature detector, 63
    olfactory, 83–84
    sensory, 66
    sleep, 181
    sound encoding, 117–120
Neurophysiology, cellular, 43–53
Neurotransmitter
    amino acid, 36–37
    inhibitory, 37
Nicotinic receptor, 42
Night blindness, 96
Night terror, 182
Nightmares, 182
Nociceptor, 72–75
Nodes of Ranvier, action potential and, 28
Noise, hearing loss and, 121
Nonadapting receptor, 60
Norepinephrine, 34
    postganglionic response, 42
Nuclear bag fiber, 137
Nuclear chain fiber, 137
Nystagmus, 153

Occipital cortex lesion, 185
Odor, types, 84
Off-surround receptive field, 96–97
Ohm's law
    cell membrane and, 11–14
    transference equation and, 14
Olfactorometer, 85
Olfactory bulb, 83
Olfactory epithelium, 83

Olfactory nerve, 83
Olfactory pathway, 83
Olfactory receptor, 83–84
Olfactory sense. See Smell
Olfactory vesicle, 83
On-surround receptive field, 97
Ophthalmoscope, 107
Opsin, 95
Optic disk, 87–88
Optic radiation, 99
Optokinetic reflex, 154
Organ of Corti, 113–114
Osseous spiral lamina, 113
Ossicular chain, middle ear, 110
Otitis media, 120
Otoconia, 151
Otolith membrane, 151
Otolith organ, 151
Otolithic reflex, 154–155

Pacinian corpuscle
    adaptation in, 60–61
    structure, 66–67
    in vibration detection, 67–68
Pain
    chronic, 71, 73
    descending inhibition, 75
    dual sensations, 72
    fiber, 72–73
    gate-control theory, 75
    inhibition, 75
    originating within nervous system, 74–75
    pathways, 72–74
    receptors. See Pain receptors
    referred, 74
    temperature and, 77
    transcutaneous electrical stimulation for, 75
    withdrawal reflex in, 139–140
Pain receptors, 71–75
    activation, 72
    function, 71
Paleocerebellum, 163–164
Paleospinothalamic tract, 73
Paradoxical sleep, 179–180
Parallel elastic component (PEC), 49, 50
Parallel fiber, of cerebellum, 165
Paralysis
    of cataplexy, 182
    skeletal muscle, 34
Parietal lobe, function, 185–186
Parkinson's disease, 160–161
Pattern recognition, 103–104

Perception
  adaptation, 62
  color, 104–105
  contrast, 102–103
  darkness, 101–102
  depth, 105–106
  patterns, 103–104
  sound. *See* Hearing
  touch sensation and, 71
  visual, 100–106
Perceptual adaptation, 62–63
Perilymph, 112
Peripheral nerves, electrical stimulation in pain control, 75
Perimysium, 43
Personality, frontal lobe and, 183–184
Phantom limb phenomenon, 74–75
Pharmacomechanical coupling, in smooth muscle, 52
Phasic receptor, 60
  stimulus application, 62
Pheromones, 79
Phosphodiesterase, in receptor potential, 96–97
Photon, in image formation, 89
Photopic vision, 102
Photopigment, 95–96, 104
Photoreceptor, 59
  in retina, 94–95
Pigment, epithelial, in retina, 94
Polymodal nociceptor, 72
Postganglionic fiber, in autonomic transmission, 41–42
Postrotatory nystagmus, 153
Postsynaptic inhibition, 40
Postsynaptic membrane, neuromuscular, 31
Postsynaptic potential
  excitatory, 36–37
  inhibitory, 37
Posture, instability, 159
Potassium
  depolarization, 18
  hyperpolarization, 17–18
  Ohm's law and, 11–14
  transference, 14–16, 21
Potassium channel, 10, 22
  activation of, 24
  nerve action potential and, 33
Potassium ion, electroneutrality and, 12
Presbycusis, 121
Presbyopia, 93
Presynaptic inhibition, 39–40
Presynaptic membrane, neuromuscular, 31
Propagation, action potential, 26–28

Proprioceptor, 58
Protanopia, 104
Pump
  cell membrane, 11
  sodium-potassium. *See* Sodium-potassium pump
Pupil, in dark adaptation, 102
Purkinje cell, 164–166
Putamen, 158
Pyramidal cell, 172
Pyramidal system, 128, 171

Quantum, acetylcholine, 33

Radiation, optic, 99
Rapid eye movement (REM) sleep, 179–180
  disorders, 182
Receptive field
  retinal, 96–97
  visual cortex, 99–100
Receptor, 57–63
  action potential, 60
  activation, 58–61
  adaptation, 60
  adapting, 60
  adequate stimulus, 58
  alpha, 42
  autonomic, 42
  beta, 42
  classification, 57–58
  components, 58–59
  dynamic, 60
  muscarinic, 42
  muscle, 135–139
  nicotinic, 42
  nonadapting, 60
  pain, 71–75
  phasic, 60, 62
  sensory coding and, 61–63
  spike generator, 59–60
  static, 60
  thermal, 75–77
  tonic, 60
  touch. *See* Touch receptor
  transducer, 58–59
Receptor potential, 58
  hair cell, 115–116
  photoreceptor, 95–97
  in sensory coding, 61
Red muscle, 131
Referred pain, 74
Reflex. *See specific type*
Refractory power, lens, 90–91
Renshaw cell, 141, 144–145

Resistor, cell membrane, 10–11
Resting potential, cell membrane. *See* Cell membrane, resting potential
Reticular formation, brain stem, 147–148
Reticular lamina, 114
Reticulospinal tract, 147, 171
Retina, 94–98
    anatomy, 94–95
    bipolar cell, 96
    cellular organization, 96
    cones, 94–95
    ganglion cell, 96, 98–99
      in color blindness, 104–105
      in contrast perception, 102–103
    geniculate nucleus, 97
    off-surround receptive field and, 96–97
    on-surround receptive field and, 97
    ophthalmoscopic examination, 107
    photochemical response in, 95
    photoreceptors, 94–95
    pigment epithelium, 94
    reception by, 96–98
    receptive field, 96–97
    receptor potential, 96–97
    rods, 94–95
    visual cycle, 95–96
Retinol, in visual cycle, 96
Reversal potential, 33
Rhodopsin, 95
Rigidity, decerebrate, 148–149
Rinne test, 121
Rod, 94–95
    in dark adaptation, 101–102
Rubrospinal tract, 171
Ruffini nerve ending
    in stimulus encoding, 68
    structure, 66–67

Saccule, 150–151
Sarcolemma, 43
Sarcomere, 44–45
Sarcoplasmic reticulum, 44–45
Scala media, 112
Scala tympani, 112
Scala vestibuli, 112
Schwann cell membrane, propagation, 28
Scotoma, 87–88
Scotopic vision, 102
Selectivity filter, ionic channel, 10
Semicircular canal, 149–150
Sensory coding, 61–63
    feature detectors in, 63
    labeled line, 63
    quality, 63

Sensory coding—*Continued*
    stimulus application rate, 62–63
    stimulus intensity, 61–62
Sensory homunculus, 69
Sensory neuron, classification, 66
Sensory pathway
    smell, 83
    taste, 80
    touch, 68–69
Series elastic component (SEC), in muscle contraction, 50
Serotonin, pain inhibition and, 75
Shivering, 76
Shock, spinal, 148
Size principle, motoneuron, 126, 133–134
Skeletal muscle
    activation, 45–51
    active state, 47
    vs. cardiac muscle, 52
    cross bridge, 45
    excitation-contraction coupling in, 46
    length-tension relationship in, 48–49
    parallel elastic component (PEC) in, 50
    paralysis, 34
    reflexes, 140–141
    relaxation, 49–50
    series elastic component (SEC) in, 50
    sliding filament hypothesis in, 48
    stretching, 48–50
    structure, 43–45
    summation in, 50–51
    tetanus and, 50–51
Skin, sensory perception. *See* Cutaneous sensation
Sleep, 177–182
    anatomic structures involved in, 180–181
    cycle, 180
    deprivation, 181
    developmental pattern changes in, 180
    disorders, 181–182
    dream, 179
    electroencephalogram in, 178
    paradoxical, 179–180
    pattern changes in, 180
    physiologic role, 177–178
    physiology, 180–181
    rapid eye movement (REM), 179–180
      age and, 180
      disorders, 182
    slow wave
      disorders, 182
      stages, 179
    stages, 179–180
Sleep spindle, 179

Sleepwalking, 182
Sliding filament hypothesis, of muscular contraction, 47–48
Slow muscle fiber, 131
Slow wave sleep, 179
   disorders, 182
Smell
   absence of ability to, 79
   clinical assessment, 85
   odor classification, 84
   receptors, 82–84
Smooth muscle activation, 52–53
Snellen eye chart, 101, 106–107
Sniffing, 83
Sodium
   Ohm's law and, 11–14
   transference, 14–16, 21
Sodium channel, 10, 22
   activation, 23
   inactivation, 23–24
   nerve action potential and, 33
   in visual cycle, 96
Sodium-calcium exchange process, 7
Sodium-potassium ATPase, 5
Sodium-potassium pump, 5–6
   material translocation and, 4
   membrane potential and, 155–156
   net current and, 15–16
Soma, motoneuron, 130
Somatosensory cortex, 69
Somatotropic representation, 69
Somnambulism, 182
Sound
   encoding, 117–120
   perception. *See* Hearing
Sound intensity level, 110
Sound pressure level, 110
Sound stimulus, 109–110
   decibels, 110
   intensity coding, 119
   location coding, 119–120
   measurement, 109–110
Sour taste, 81–82
Space constant
   action potential, 27
   inhibition, 38–39
Spatial summation, 38
Speech, ability deficits, 185
Spike generator, 59–60
Spinal cord
   in pain sensation, 72–73
   reflexes, 126–127, 135–145
   shock, 148

Spinal cord—*Continued*
   in touch sensation, 68–69
Spinocerebellum, 164
   lesions to, 166–167
Spinoreticulothalamic pathway, 73
Stapes, 111
Static receptor, 60
Steady state condition, 13
Stereopsis, 105
Stevens power law, in sensory coding, 61–62
Stimulus
   application rate, 62–63
   gustatory, 81
   intensity encoding, 61–62
   quality, 63
   sound. *See* Sound stimulus
   touch receptor, 67–68
Stretch reflex, 140–141
   functional, 172–173
Stria vascularis, 113
Striatum, 158–159
   lesions to, 160
Substantia nigra, 158–159
Subthalamic nucleus, lesions to, 160
Succinylcholine, 34
Summation, 38
   in muscle contraction, 50–51
Superior olive, 120
Sweating, 76
Synapse
   anatomy of, 29
   axoaxonic, 39–40
   neuromuscular, 30
Synaptic cleft, 31
Synaptic terminal, alpha motoneuron, 36
Synaptic transmission, 29–42
   chemical, 30–35
   action potential initiation and, 35
   in autonomic nervous system, 40–42
   desensitization and, 34
   end plate potential and, 32–33
   excitation-secretion coupling and, 31
   interface, 34
   miniature end plate potential and, 33–34
   myasthenia gravis and, 34–35
   neuromuscular junction and, 30–31
   postsynaptic membrane and, 31
   presynaptic membrane and, 31
   reversal potential and, 33
   synaptic vesicle cycle and, 32
   transmitter inactivation in, 34
   within central nervous system, 35–40

Synaptic transmission—*Continued*
   electrical, 29–30
   inhibition, 38–40
      postsynaptic, 40
      presynaptic, 39–40
   interference with, 34
Synaptic trough, 31
Synaptic vesicle
   cycle, 32
   neuromuscular, 31

T tubule, 44–45
   calcium release and, 46–47
   in cardiac muscle, 51
Taste
   absence of ability to, 79
   clinical assessment, 85
   pathways, 80
   receptors, 80–82
   sensations, 81–82
   transducer in, 59
Taste bud, 80–81
Taste cell, 80
Tectorial membrane, 113, 114
Temperature, sensation, 75–77
Temporal lobe, function, 184–185
Temporal summation, 38
Tendon organ, 136, 144–145
TENS (Transcutaneous electrical stimulation), 75
Terminal cisterna, 44–45
Tetanus, 51
Thalamus, in pain sensation, 73
Thermoreceptor, 59, 75–77
Thermoregulation, REM sleep and, 180
Tinnitus, 121
Tongue, 80–82
Tonic neck reflex, 154–155
Tonic receptor, 60
Tonometry, 107
Topographic representation, 69
Touch receptor, 65–71
   free nerve ending, 66
   information synthesis, 71
   Meissner corpuscles and, 66–67
   Merkel's disks and, 67
   neuron classification, 66
   Pacinian corpuscles and, 66
   Ruffini nerve endings and, 66
   sensory pathway, 68–69
   stimulus application rate encoding, 68
   stimulus intensity and location and, 68
   topographic organization, 69–70

Touch receptor—*Continued*
   two-point discrimination, 70
   vibration encoding, 67–68
Touch spot, 67
Transcutaneous electrical stimulation (TENS), 75
Transducer membrane, 58–59
Transducin, in photoreception, 95–96
Transference, 14–15
Transference equation, 14–16
Tremors, 161
   ataxia, 167
   intention, 167
Trigger calcium hypothesis, 47
Tropomyosin, 45–46
Troponin, 46
Tuning fork, 121
Two-point discrimination, in touch sensation, 69

Unmyelinated axon, propagation in, 28
Utricle, 150–151

Vestibular apparatus, 149–151
   function, 152
Vestibular system, in motor control, 126, 127
Vestibulocerebellum, 163
Vestibulomotor reflex, 153–154
Vestibuloocular reflex, 153
Vibration
   basilar membrane, 114–115, 117
   detection, 67–68
Vision, 87–107
   anomalous color, 104
   binocular, 105
   clinical assessment, 106–107
   color, 104
   correction, 93–94
   double, 102
   eye structure, 88–89
   far point, 93
   image formation, 89–94
   mesopic, 102
   near point, 93, 107
   perception, 100–106
   photopic, 102
   retina, 94–98
   scotopic, 102
   transducer in, 59
   visual cortex, 98–100
Visual acuity, measurement, 106–107
Visual cortex, 98–100

Visual cortex—*Continued*
   cortical columns, 100
   in depth perception, 105
   pattern recognition, 103–104
   receptive field organization, 99–100
Visual cycle, 95–96
Visual perception, 100–106
   color blindness, 104–105
   color vision, 104
   contrast perception, 102–103
   dark adaptation, 101–102
   depth perception, 105–106
   pattern recognition, 103–104
   visual acuity, 101, 106–107
Vitamin A, in visual cycle, 96
Vitreous humor, 88, 89
Volley code, for sound frequency, 118–119

W ganglion cell, 99
Wakefulness
   ascending reticular activating system and, 179, 180–181

Wakefulness—*Continued*
   disorders, 181–182
Warm fiber, 76, 77
Wave, sound, 109–110
Weber test, 121
Weber-Fechner law, in sensory coding, 61–62
Wernicke's aphasia, 185
White muscle, 131
Withdrawal reflex, 139–140

X ganglion cell, 98–99

Y ganglion cell, 99

Z band, myofibril, 44–45